LOMA VISTA
P.T.A. LIBRARY

DISCARDED

DATE DUE

76859

591.5 Barker, Will
Bar   Winter-sleeping
          wildlife
                                    B 1-171

Winter-sleeping wildlife /
591.5 Bar   16646

Barker, Will.
        LOMA VISTA SCHOOL LIBRARY

WINTER-SLEEPING WILDLIFE

Illustrated by

CARL BURGER

HARPER & ROW, PUBLISHERS, NEW YORK AND EVANSTON

WINTER-SLEEPING WILDLIFE

by WILL BARKER

Foreword by Ernest F. Swift,
Executive Director,
National Wildlife Federation

WINTER-SLEEPING WILDLIFE

Copyright © 1958 by Will Barker

Printed in the United States of America

All rights in this book are reserved.
No part of the book may be used or reproduced in any manner whatsoever without written permission except in the case of brief quotations embodied in critical articles and reviews. For information address:
    Harper & Row, Publishers, Incorporated,
    49 East 33rd Street, New York 16, N. Y.

Library of Congress catalog card number: 58–5298

For
>   GINNY, NICK, and JO,
>   as lively as crickets.

Foreword

Winter-Sleeping Wildlife presents a fascinating phenomenon in the world of certain North American animals.

This phenomenon is hibernation, a mysterious cycle of life, a way of survival while the world is cold and harsh. It is a function that carries on in a world of its own, oblivious to atomic mankind.

It may be just a little world under a clod by the garden gate, or it may lie beneath the bark of a sickly tree fighting for existence on a city street. It may be a hidden world beneath a farm hedgerow, or a world in the primeval fastness of a mountain wilderness. The world of hibernation is all around us, an annual miracle of nature.

Although man in his effort to reach the planets may grow inattentive and indifferent to the everyday miracles of nature which surround him, he cannot stand apart. He cannot escape his environment even while amusing himself with his push-button world. This life around him has a profound effect upon his conduct; his own life depends completely upon the world of nature, even though its bounty comes to him third-hand in tin cans and frozen packets.

In *Winter-Sleeping Wildlife*, Will Barker raises the cover to give you a look at one of the mysteries of this world of nature. He presents his subject for the young and old, as well as for the casual conservationist and the wildlife professional.

The questioning youngster who wonders why some creatures simply disappear in the cold months finds his answer here. He learns that he can pursue the contemplations of nature in his own back yard, or the city park.

This book was written with the wondering youth in mind. And yet it is not written as a child's book. *Winter-Sleeping Wildlife* is written for those who want to understand a mysterious phase of the world they live in. It will invite the adult to return to a more leisurely world, when he had time for more than blinking traffic lights and the jangle of telephone bells. Even the professional wildlife conservator is likely to delight in this fresh-eyed look at the fascinating byways of the world he works in.

>Ernest F. Swift, Executive Director
>National Wildlife Federation
>Washington, D. C.

Acknowledgments

For the information furnished me while writing *Winter-Sleeping Wildlife*, my publishers and I wish to thank all those who have given of their time, energy, and knowledge.

We also wish to thank the following organizations: United States National Museum; American Museum of Natural History; National Wildlife Federation; Sport Fishing Institute; Bureau of Entomology and Plant Quarantine and United States Forest Service, United States Department of Agriculture; and United States Fish and Wildlife Service and National Park Service, United States Department of the Interior.

If I have been mistaken in interpreting any of the information so graciously furnished me the fault is all mine.

To Carl Burger, illustrator for this and for *Familiar Animals of America*, my special thanks for his invitation to sit down with him in his Pleasantville studio and plan illustrations for *Winter-Sleeping Wildlife*. Illustrations for both books are of the quality of those by Louis Agassiz Fuertes, the American wildlife painter under whom Carl Burger studied.

And to Ursula Nordstrom, my Harper editor for this book, and her assistant, Susan Carr, I wish to express my appreciation for the painstaking care with which they checked and double-checked the typescript for this book.

W. B.

Washington, D. C., 1958

Contents

LIST OF ILLUSTRATIONS	xiii
INTRODUCTION	1
MAMMALS	4
Bat	5
Woodchuck	10
Prairie dog	18
Ground squirrel	23
Eastern chipmunk	28
Jumping mouse	33
Bear	37
Badger	45
Raccoon	49
Skunk	53
BIRDS	59
Poor-will	60
Hummingbird	63
REPTILES	65
Garter snake	65
Box turtle	69
Snapping turtle	74

AMPHIBIANS	79
Toad	79
Spring peeper	83
FISHES	86
Carp	86
Mackerel	90
Basking shark	90
SPIDERS	93
Black and yellow garden spider	94
INSECTS	100
Lady beetle	101
Cabbage butterfly	105
Woolly bear	107
Grasshopper	110
Praying mantis	113
Dragonfly	116
Bumblebee	120
MOLLUSKS	126
Garden snail	126
IN CONCLUSION	130
LIST OF COMMON AND SCIENTIFIC NAMES	135

Illustrations

Little brown bat hibernating	6
Plan and cross-section of woodchuck burrow; woodchuck in hibernating position	13
Hoary marmot, yellow-bellied marmot, and woodchuck coming out of winter quarters	15
Cross-section of a prairie-dog burrow	19
Black-tailed prairie dog	22
Ground squirrel	24
Chipmunk in hibernation; cross-section through burrow	30
Jumping mouse	34
Jumping mouse in hibernating position	37
Black bear in winter den	38
Black bear emerging from winter sleep	43
Badger	46
Raccoon leaving den to look for food	51

Striped skunk about to enter burrow where others are already asleep	54, 55
Poor-will	61
Ruby-throated hummingbird	63
Garter snakes congregating to hibernate	67
Box turtle	70
Snapping turtle	77
Toad in hibernating position	80
Spring peeper peeping	84
Carp burrowing into mud at bottom of pond	87
Basking shark and mackerel	91
Black and yellow garden spider; cocoon in which young pass the winter	95
Lady beetle attacking an aphid; larva and pupa of lady beetle	104
Cabbage white butterfly	106
Woolly bear and Isabella tiger moth	109
Grasshopper depositing eggs; young grasshopper	111
Praying mantis	114
Adult and nymph dragonfly	118
Bumblebee	123
Garden snail	128

WINTER-SLEEPING WILDLIFE

Introduction

WHEN WINTER comes, the animals of North America have different ways of meeting this seasonal change in living conditions. One way in which some animals react to winter is by passing a part or all of this season in deep sleep.

Profound and insensible, this sleep is known as hibernation. This word comes from the Latin. It means "to winter" or "to pass the winter in close quarters, in a torpid or lethargic state."

No one knows exactly how hibernation came into being. Probably it originated because certain animals needed protection from the cold or had to avoid a season when food was scarce. And also, like migration, it undoubtedly has a connection with reproduction. For the glands of the deep-sleeping animal greatly increase in development once hibernation is over. This is a decided contrast to the condition of the glands in the fall. Then they are not so fully developed and as a result they are least active.

Only animals that can store up enough food resources in the form of accumulated body fats are able to hibernate.

These deposits of fat may provide the stimulus that induces an animal to hibernate—rather than merely the external conditions of cold weather and lack of food.

In addition to its general winter fat, a hibernating animal has other special fats. These are in the form of dark-colored deposits and tissues. They form around the blood vessels in the neck, chest, and other parts of the body. During hibernation such fatty areas grow smaller, but they shrink at a much less rapid rate than other fats.

Hibernation has its summer counterpart—a sleep that is known as estivation. This word also comes from the Latin and means "to pass the summer in a state of torpor." But among North American animals there are only a few that estivate.

Some of our animals are not true hibernators, but they do pass a part or most of the winter asleep. At the first indication of really cold weather or during storms these animals hole up in protected spots. Unlike hibernating or estivating animals, the body functions of these "catnapping" species do not slow until the animal is in a deathlike state.

Insects undergo a special type of hiberation. This is known as the *diapause*, a temporary stoppage of activity or growth at an immature stage.

As a rule hibernating animals, particularly mammals, have longer lives than many of those that are active at all seasons. The common shrew is active for the better part of each of the year's 365 days. This constantly busy little animal, weighing ⅓ to ⅘ of an ounce, wears itself out in eighteen months.

A hibernating bat of about the same weight as the shrew may live seven or eight years. The much longer life span of the bat is the result of the yearly intervals of deep sleep. These intervals give the body machinery a complete rest.

Not a great deal is known about hibernation, but the medical profession is studying this phase of animal life. If investigators are able to learn how a woodchuck can live with its body temperature at 37° F., then they may be able to improve the technique of "deep freezing" human beings for operations on the heart.

To date the temperature of a human being cannot be reduced with safety to more than 20° F. below the normal temperature of 98.6° F. But if and when the whys and wherefores of hibernation are fully understood by science, then man will benefit in still another way from the animals around him.

Mammals

MAMMALS ARE the highest class of backboned or vertebrate animals. The name "mammal" comes from the Latin *mamma*, a word meaning breast. A female belonging to this class has glands that secrete milk to nourish the young after birth.

With the exception of the egg-laying mammals of Australia, the young of mammals are born alive and helpless. The young of the opossum are the most helpless of all. They are undeveloped at birth and have to live in the pouch of the mother until they are strong enough to withstand the outside world.

Mammals are warm-blooded and usually have a body partly or wholly covered by hair or fur. Most of them have four legs, though the legs of the bat and those of the seal have undergone changes to fit these two mammals for the type of life each leads. And the hind legs of a number of marine mammals are present in the body but so little developed that they do not even show.

Though mammals are the most outstanding form of life on earth, there are not so many of them as there are other species of animal. In fact there are only about 3,500 known mammals in the world. Of this number some 1,500 live in America north of the Rio Grande.

Only mammals of North America that sleep a part or all of the winter are written about here. They fall into two categories: those that are the deep sleepers, the hibernators, and those whose sleep is less profound or even fitful, the "catnappers."

BAT

One of North America's hibernating mammals is most unusual. This unique hibernator is a bat—the only mammal of all the world's mammals that can fly.

North America has many of the world's 2,000 known and named bats. There are eight bat families on this continent. In one of these families is the "little brown bat." This tiny pug-nosed creature inhabits most forested areas from the North to the South and from the East to the West.

As a rule the little brown bat weighs no more than $\frac{1}{7}$ to $\frac{1}{3}$ of an ounce. But by the time it is ready to hibernate, this flying mite has doubled in weight. To achieve this increase the bat eats just as much as it can during the four to six weeks before the time comes to settle down for winter.

Usually the little brown bat selects a cave in which to pass the winter. Suitable, too, are hollows of trees, the undersides of eaves of houses, or empty buildings. These are but a few of the many places in which this bat hibernates. In any of these spots it hangs up by one foot, then another, or perhaps all four.

The cave or other hibernating spot has to have a temperature of no less than 30° and no more than 40° F. above

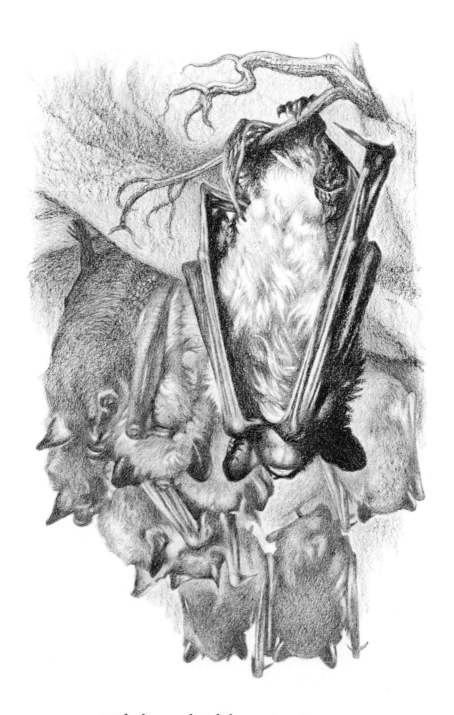

Little brown bat hibernating in a cave.

zero. If the temperature is within this range, the bat goes slowly to sleep. The breathing grows less and less frequent. Eventually the animal takes only one breath in every five minutes.

The body temperature, or blood heat, drops until it is only a little higher than the surrounding air. Sometimes it is no more than 1° higher.

The lowering of the blood heat reduces the chemical activity of the bat's body to match the available energy. The source of the bat's energy in winter is the fat accumulated during its enormous prehibernation feeding. By the time spring has come, the bat weighs about one-third less than it did upon entering hibernation.

If the hibernating spot is too warm, the chemical activity of the bat's body continues at too high a rate. Consumption of the stored fats is then too great also. As a result a bat is apt to starve.

If the temperature of the hibernating spot drops too suddenly, the little mammal may freeze to death. A gradual lowering of the temperature apparently sounds an alarm of some sort. When this happens, the bat usually flies to another spot in which the desired temperature is constant.

If the little brown bat is disturbed during hibernation it rouses enough to fly to a new roost. And if there is unfrozen water, the animal may take a sip or two. Upon emerging from hibernation the little brown bat shivers as it comes slowly back to its active state.

Some of these bats mate during hibernation. Others mate in late summer or early fall. Young bats may be born from early spring until midsummer. They are usually born in the cave in which the mother hibernates. Less frequently a birth occurs while the mother hangs from a tree.

If it is a cave birth, the female flies to a special section where other females have gathered to bear young. In this "maternity ward," the female hangs up by her feet and thumbs, with the head toward the floor of the cave. She stretches wide her flaps of leathery skin. Known as membranes, these flaps run along the body from the forelegs to the hind legs, then back to the tail.

When the membranes are pulled taut they form a sort of hammock. This is used somewhat in the manner of a fireman's net into which people jump from burning buildings. As the baby bat emerges during the birth process, the tiny creature drops into the membranes.

A newly born bat is a large baby in relation to the mother. The young animal is one-eighth to one-third the size of its parent, who measures $3\frac{1}{2}$ to $4\frac{1}{2}$ inches in length.

As soon as it is born the little bat clings to its mother's breast. This is its haven for the next two weeks. She helps it stay in place by partly folding her membranes around it. The baby bat nurses at one of two nipples.

At first the mother carries her baby each time she flies out in the evening. But at the end of two weeks the young bat has gained so much weight that she cannot carry it. She leaves it at home. It hangs itself up to await her return. By the time a young bat is three weeks old, it starts practice flights. And in no time at all it learns to catch its own food.

The little brown bat drinks and feeds when it is in flight. To drink it flies low over the surface of any convenient water. On each down dip it laps up a few drops. Down-dipping is repeated until the animal has had enough to drink.

The insects on which the bat feeds are caught in the mouth or the tail membrane. By lowering the tail membrane a scoop is formed. This catches the insects while the bat flies along in search of its evening meals.

The little brown bat is always a big eater in relation to its size and weight. At any one time the amount eaten may equal as much as one-fourth of the bat's weight. But before the night ends the little flier frequently eats so many insects that the food ingested equals one-half its weight.

As the little brown bat flutters through the evening dusk it guides itself by locating an echo. This means of charting a course is known as "sonar." Every second the bat is in flight it emits a series of supersonic squeaks. The vibrations caused by these squeaks range in frequency from 30,000 to 70,000 a second. The average frequency is about 50,000 a second. At this pitch the duration of a squeak is slightly less than $2/100$ of a second.

These vibrations are of such high frequency that you and I are unable to hear them. Our hearing is limited to a range of 16 to not quite 30,000 vibrations a second.

Even at rest a bat emits a supersonic squeak of low frequency. But the instant it starts to fly, the bat begins to "broadcast" at an increased rate. At first the squeak rate is not so great. But as an obstacle is approached the bat increases the number of squeaks a second. Consequently the pitch of the returning vibrations becomes higher and higher. Near an obstacle the rate is at its peak. But as hazards are passed the rate steadily diminishes.

Broadcast sounds are turned aside by any object in the path of flight. Echoes then bounce back to the bat, whose unusually sensitive ears pick up the returned sound. The ability of the bat to pick up the echoes is possible because muscles in the ears close for a moment. This closure shuts out the squeaks still being broadcast.

Prehibernation flights are generally over by the middle of September. Then the little brown bat seeks the safety of cave, tree hollow, abandoned building, or other spot where the temperature and humidity are such that it can overwinter in safety.

In such spots this bat hangs itself up to sleep away the better part of the winter. Sometimes it sleeps alone. More frequently it joins other species of hibernating bats.

The big brown bat, the little brown bat, and the pygmy bat all winter together in an abandoned New Jersey iron mine. Another species occasionally found hibernating in mixed company is the pale, yellowish-brown pipistrelle. This is our smallest bat; it measures only $2\tfrac{2}{5}$ to $3\tfrac{2}{5}$ inches in length.

The pipistrelle hibernates from the middle of October until the middle of March. But it does not carry on an uninterrupted sleep. This bat frequently changes its roost. And on warm days in winter it comes out and flies around.

After a short flight the pipistrelle returns to the cave. It hangs itself up, either alone or in groups that seldom number more than 50. Now and then these bats gather by the hundreds in one spot for the purpose of hibernating.

A cave in Indiana was the wintering quarters of 491 eastern pipistrelles. They were jammed into a space 1 foot wide by 1 and $\tfrac{7}{10}$ feet long.

So, if you wish to study bats in great numbers, go to a cave in winter. In such a spot at this time of year, you will usually discover fast asleep some members of the world's only order of flying mammals.

WOODCHUCK

By fall the flat-headed woodchuck looks as if it might burst its coat of yellow-brown fur. This fat-lady look is due to constant daily gorging on plant foods all during the season when wild and cultivated crops are ripe and abundant.

Known, too, as the groundhog and the whistle-pig, the short-eared woodchuck comes out of its underground burrow to feed three times a day. You can often see one early in the morning, late in the afternoon, or again early in the evening. These are the usual mealtime hours. But at summer's end the 'chuck eats as if its stomach were a bottomless pit that never could be filled.

A woodchuck rears up on its short hind legs to eat, then chews off the flowers, leaves, and soft stem parts of many different wild plants. These include clovers, dandelions, and buttercups.

Farm and garden crops are also a part of the diet. The 'chuck has a particular liking for grains. Some feedings include as much as 1½ pounds of alfalfa, barley, or corn. This enormous eating in late summer or early fall is to acquire thick layers of fat. Plenty of fat is necessary to sustain the woodchuck while it hibernates.

Shortly before going underground for the last time of the year the woodchuck stops all eating. The indifference to food lasts ten days or two weeks. During this interval the 'chuck spends a part of each day dozing in the sun at the burrow entrance. At other times it wanders around near home, stopping, perhaps, to gnaw halfheartedly on something, but never really eating. The prehibernation period is marked by irritation, if the behavior of a wild 'chuck is similar to that of a tame one.

Finally the day comes when it is time to hibernate!

Then, fat to the point of bursting, the 'chuck waddles into its burrow system of many connecting passageways and chambers. It shuffles along one of the dark tunnels to the highest and largest chamber. Lower rooms are generally the smallest, and one of them is used solely as a toilet. Some upper chambers measure 15 to 18 inches in diameter and are often 7 to 8 inches in height at the ends.

The woodchuck usually selects the highest room in which to pass the winter. Such a spot is less likely to be flooded than chambers at lower levels. Here the animal snuggles down in a bed of dried grasses. This material was cut and carried into the burrow during the summer. Occasionally a 'chuck winters in one of the unlined chambers along the sides of tunnels.

The entrance to the winter sleeping chamber is closed by dirt. The 'chuck scrapes the dirt from the far end of the room. Sealing the door is a safety measure. A closed door keeps out opossums, skunks, or rattlesnakes.

These creatures frequently try to share the winter bedroom of a 'chuck for varying lengths of time. The warm-blooded 'possums and skunks seek a woodchuck's burrow to drowse away days that are extremely cold or snowy. But the cold-blooded rattlesnakes often move right in to pass the entire winter.

Once the entrance is blocked, the woodchuck rolls up into a ball by putting its head down between the short hind legs. The snubbed nose touches the more lightly colored fur of the belly, while the tail of 5 or 6 inches is brought up over the head and down the spine. The eyelids and the lips are squeezed as tightly together as possible.

Curled thus in its bed of straw or on the dirt floor of a side chamber, the woodchuck falls gradually asleep. Eventually, however, the sleep becomes deep.

When a woodchuck is active, it breathes about 2,100 times an hour. When it becomes excited, the rate skyrockets to as much as 6,000 times in sixty minutes. But when it is in deep sleep, the rate decreases to such an extent that

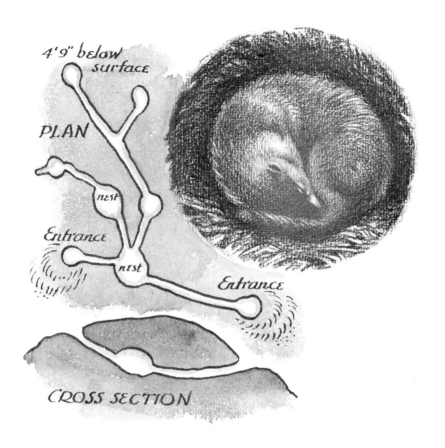

Plan and cross-section of woodchuck burrow.
INSET: Woodchuck in hibernating position.

the number of times the 'chuck breathes in an hour is only 10!

The rhythmical beating of the pulse becomes less frequent. In fact the pulse beat grows so faint that it is difficult to detect. The animal grows colder and colder. The normal temperature of about 100° F. drops lower and lower. Finally it is somewhere between 57° and 40° F.

Now the woodchuck is actually hibernating. So deathlike is the sleep that if you should touch the animal or blow in its ear, it would not rouse.

How long a woodchuck hibernates depends upon the part of North America in which it lives. Woodchuck territory extends from northeastern Quebec and Nova Scotia across the continent to southwestern Yukon.

In the East these animals live as far south as eastern Virginia and northern Alabama. And in the West the home of the woodchuck is as far south as western Montana, eastern South Dakota, and northeastern Oklahoma.

On the most northerly part of the range the woodchuck often settles down for the winter by the middle of September. Hibernation may last for at least six months. The length of hibernation for a southern 'chuck is much shorter. But the animal's two cousins, the hoary marmot and the yellow-bellied marmot, often sleep much longer. Sometimes these two animals hibernate for seven and one-half months.

The hoary marmot and the yellow-bellied marmot live at high altitudes in the mountain ranges of the West and the Northwest. In these regions, winters are severe and snows are deep.

Often when the woodchuck's two cousins wake up in the spring the ground is blanketed by as much as ten feet of snow. The awakening of marmots is governed by the alti-

Three cousins coming out of winter quarters.
TOP: Hoary marmot lives in high altitudes at timber line.
MIDDLE: Yellow-bellied marmot lives below timber line.
BOTTOM: Woodchuck lives in lowland meadows.

tude and the latitude—the elevation from sea level and the distance in degrees from the equator.

Usually, the yellow-bellied marmot is the earlier spring riser of the two. This animal lives at lower mountainside levels than the hoary marmot, whose home is apt to be at timber line. This is the area on a mountain that marks the end of tree growth.

If you enjoy climbing in the mountains of the West, you might see a yellow-bellied marmot. This orange-colored mammal likes to sun-bathe. It settles down on outcroppings of rock in deep mountain valleys or on convenient ledges as far up a mountain as 10,000 feet.

To see the grayish hoary marmot, you have to climb high in the mountain ranges of northwestern Alaska and the Alaska Peninsula or in the mountains of central Washington and Idaho.

The silences of such places are broken by the clear, shrill whistles of this marmot. Due to these whistles the hoary marmot was called "the whistler" by fur trappers in the Rocky Mountains. Occasionally the woodchuck is called the whistler, too. But this name is not appropriate, for the woodchuck is a chirper.

When the woodchuck and its marmot cousins come out of their burrows in the spring they look like shadow-shapes. The animals are so thin that their coats seem several sizes too large and hang loosely on their bodies.

A fully grown woodchuck weighs 4 to 10 pounds, and extra-large ones weigh as much as 14 pounds. During hibernation a 'chuck loses one-third to one-half its weight. So you can well imagine how skinny these animals are in the spring.

After such a long fast you might think a 'chuck would be so hungry that it would eat as soon as it could. But this is not true. No matter how hungry the animal might be, the first thing it does is to seek a mate.

A male roams around on its own territory or that of another male, looking for a mate. In its travels a male usually meets other males trying to find a mate. When their paths cross, males fight.

Fights are noisy affairs. The 'chucks squeal and growl and snap. They use their sharp white incisors as weapons to chew opponents' hides, ears, and tails. Occasionally as much as half of a 6-inch tail is nipped off. Sooner or later, the weaker of the fighters quits. The winning 'chuck takes a mate, while the loser wanders away to continue his search.

Some males have only one mate. Others have several, one after another. As soon as the mating season is over, males return to their own territories and start to eat.

About a month later 2 to 6 young 'chucks are born. They stay with their mother until midsummer. Then she drives them away from her burrow.

They find temporary homes in nearby abandoned burrows. The mother continues to watch out for her scattered family during its first weeks away from home. When danger threatens, she chirps a warning. This beware call sends the small animals scuttling to the safety of their first homes.

Soon each young 'chuck moves on to a territory of its own. Though the woodchuck was formerly a forest animal, it now seems to prefer a bushy, wooded area bordering a meadow in which there is plenty of clover.

By the time fall comes the partly grown 'chuck has dug a short but adequate tunnel on the territory of its choice.

Here, after conditioning itself for hibernation, first by feasting and then by fasting, the young animal curls up into a ball to pass the first winter of its four-to-five-year life.

PRAIRIE DOG

To see the fat, short-tailed prairie dog in natural surroundings, you have to be in western North America. Out on the Great Plains, up in the Rocky Mountains, or down in northern Mexico, this buff-colored animal lives in one of the most elaborate homes made by any American mammal.

The home of the prairie dog is a maze of connecting tunnels. These lead to rooms at various levels. Rooms are at the ends of vertical tunnels, and sometimes one may be 16 feet beneath the earth's surface.

A room at this depth is perfect for hibernating. But not all prairie dogs hibernate! Some are active throughout the year. Others go below ground during extremely cold weather, only to come out on balmy days. And still others pass the winter in deep sleep. The prairie dogs on the northern end of the range or those living at high altitudes are usually the animals that actually hibernate.

If you visit prairie-dog country you will learn that there are two species of this lively little animal. In short-grass country you usually see the black-tailed prairie dog. The tail of this stocky animal is more than one-fifth of its total length of 12 to 16 inches.

But in grassy uplands or in mountain regions you are apt to see the white-tailed prairie dog. Smaller and more slender than the animal of the short-grass country, this prairie dog has a white-tipped tail. It is less than one-fifth of the total body length of 12 to 14½ inches. And there are dark spots above each orange-colored eye and on each cheek.

Both these prairie dogs live in colonies known as "vil-

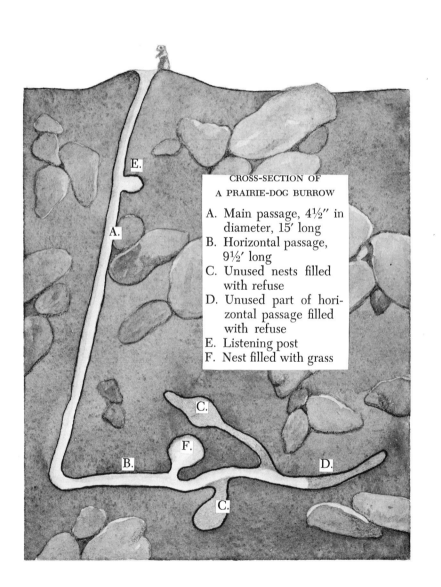

lages." Up until 1860 there were many villages. Some of them were so large that they covered many square miles. But after the Civil War the expansion of the country to the West and the demands of an ever-growing population encroached upon prairie-dog habitat. Today, only a few villages remain.

Frequently the mountain villages of the white-tailed prairie dog are snowbound for weeks at a time. The "dogs" in these villages often hibernate by the middle of October and seldom are they seen after Thanksgiving.

All summer long both species of prairie dog eat as much as they can. The one with the black tail devours almost any plant within easy reach of its tunnel entrance. The diet includes such grasses as blue, grama, and wheat. And if there are plenty of grasshoppers around, the animal gobbles down these insects, too.

The white-tailed prairie dog eats shrubby plants in addition to various weeds and grasses near its home. During summer and early fall it adds grubs and adult beetles, grasshoppers, and the larvae of butterflies to the diet.

This heavy day-in-and-day-out eating readies the prairie dog for hibernation long before the time comes to go to sleep. When it is so fat that it waddles, the prairie dog enters its home for the last time of the year. The entrance to the vertical tunnel of the black-tailed species is protected by a dirt mound in which there is a hollow center. This mound is similar in appearance to the crater of a volcano. It is thrown up to keep the tunnel from being flooded during heavy rains.

The entrance to the home of the white-tailed prairie dog has a low mound or none at all, for the burrow of this species is in high country, where rain water runs off easily.

A few feet below the entrance there is a small side

chamber. This is the listening post, to which a prairie dog scuttles in times of danger.

The presence of an enemy in a village is announced by a prairie-dog sentry. This animal rears up on its haunches and whistles loudly.

Some people believe that the warning sounds like "Skip!" "Skip!" "Skip!" But whatever the sentry's warning seems to say, it is enough to make every dog in the village dash for the safety of its burrow.

Each animal waits at its listening post for the sentry to signal by another whistle that the danger is past. If the safety whistle is not sounded, each waiting dog scurries farther down in its tunnel. Here it remains as long as is necessary.

The room in which a prairie dog hibernates is far beneath the listening post. This chamber is lined with dried grasses—the bedclothes of the hibernating animal.

For its winter sleep the prairie dog rolls into a ball. The head is bent down so that the chin touches the lighter-colored fur of the belly. The eyes and mouth are closed tightly, and the ears are folded back and laid as flat against the head as possible.

To complete its hibernating position the prairie dog brings its hind feet forward and holds them on each side of the nose. The tightly clenched forefeet either are held under the chin or along each side of the cheeks. And the tail is tucked between the legs, with the tip covering the nose.

In this compact position the warm-blooded prairie dog becomes cold-blooded. The body temperature decreases and so does the animal's breathing. Unless there is a great change in the temperature of the hibernating chamber, a dog will sleep for some time without once waking up.

A near miss. Black-tailed prairie dog and Swainson's hawk.

The length of hibernation varies, and is, of course, influenced by altitude and latitude.

For either the hibernating or non-hibernating prairie dog the mating season is February, March, or April. A month later the female bears 4 to 8 pups in a deep, grass-lined chamber.

When a pup is four weeks old, it comes out of the burrow to frisk around the entrance and to nibble on any nearby plants. And when the summer ends, a young dog is able to take care of itself.

The prairie dog has a number of enemies. The badger, the bobcat, the coyote, and the hawk and the eagle prey on it. But if the now-scarce prairie dog escapes these and other predators, it may hibernate each winter for at least eight years, or even for as many times as one captive dog. This "Methuselah" went to sleep winter after winter for eleven years.

GROUND SQUIRREL

A ground squirrel once slept for thirty-three weeks without waking up for even a few minutes. Many of these earth-bound squirrels pass at least three-fourths of their lives asleep in circular underground rooms. These chambers are lined with soft, well-dried grasses.

Such deep and prolonged sleep is the rule for most ground squirrels wherever they live—in Europe, Asia, Africa, or North America.

On this continent we have thirty-one species of ground squirrel. You will find these little animals—somewhat like chipmunks in appearance but lacking face stripes—in most of western North America. They live in all kinds of habitats, from the tundra edging Bering Sea and south all the way to the deserts of Mexico.

A ground squirrel may sleep eight months

The largest and one of the most numerous of our ground squirrels is also one of the longest sleepers. This gray or gray-brown animal is the Columbian ground squirrel, a bushy-tailed member of a large family in which there are species with short-haired tails.

By July this squirrel has already acquired its heavy winter coat. And in July it retires to a special sleeping den, where it remains until the following February or March.

A rocky slope with trees seems to be the favorite home of the Columbian ground squirrel. But the animal lives in lowland country and mountain valleys, too. Any area must have plenty of growing plants. This squirrel drinks no water. Therefore, it depends solely on plant juices to furnish the liquids it needs in order to live.

As soon as plants start to grow in the spring, the Columbian ground squirrel begins to eat as much as it can. It gorges on all sorts of cultivated crops. And it also eats the leaves and stems of many wild plants, the flowers of buttercups and dandelions, and the berries of the currant and the strawberry.

The squirrel stuffs itself so that thick layers of fat build up on the body. Accumulated fat is the source of energy that sustains the sleeping squirrel for seven or eight months. There must also be enough fat left in the spring to keep the animal alive for two weeks after it comes out of hibernation. This is at a season when growing plants are scarce.

To get ready for winter the squirrel has to "step lively" during its short, active period aboveground. It hollows out a special sleeping den in a burrow system of numerous tunnels and many rooms. The circular den is 6 inches to 5 feet below the surface of the earth, and its diameter is 9 to 10 inches.

To make a comfortable bed the squirrel lines the den

with dried grasses. And to provide an easy exit when it wakes up in the spring, the animal digs a tunnel to within a short distance of the surface.

By the time this work is completed the ground squirrel is so fat that its bulging little belly almost touches the ground. By this time, too, green foods are scarce or wanting due to the lack of rains. And the heat is so intense that the squirrel is uncomfortable.

As the hot, dry days continue the ground squirrel eats less and less and becomes more and more drowsy. Finally the animal drags itself to the specially prepared sleeping cell. It "shuts the door" behind it by using its nose to pack loose earth into the opening.

Once the last particle of dirt has been rammed into place, the now half-asleep squirrel is ready to hibernate. It bends backward until it sits on the hips. Then the head and body are arced forward until the nose presses against the full, rounded belly.

Now the animal has shaped itself into a sort of hoop. This position presses out most of the air in the lungs. In a short time the breathing and heartbeat slow almost to the stopping point. The body temperature drops from a normal 90° F. to 40° F. With body functions reduced to a minimum, the animal lapses into a deathlike sleep.

If the sleeping squirrel's temperature falls below 40° F., the animal stirs. Apparently this movement increases breathing, heart action, and circulation. This results in a slight rise in temperature, which, in turn, prevents the squirrel from freezing to death. Once the danger of dying has been overcome the animal drops back into deep sleep.

But if the squirrel fails to rouse, it freezes to death!

As the ground squirrel goes to sleep in midsummer and

continues sleeping until late winter or early spring, the animal both estivates and hibernates.

The first of these squirrels to rouse from this summer-winter sleep are old males. They start up the partly completed exits, prepared the previous summer, then burrow through the remaining earth to the surface. Ten days later the females appear. And last of all come the young.

Most ground squirrels are extremely hungry at this time of year. But there is hardly any food to be had, and none in quantities will be available until the spring rains moisten the earth. Then, there will be such wild bulbs as those of the camas, onion, and glacier lily.

Some old males, however, do not have to wait around with empty stomachs for these foods, because the year before, when food was plentiful, they laid in supplies.

These seemingly foresighted squirrels carry seeds and bulbs to their burrows. The foods are stored in small holes beneath the sleeping den. And then, when there is nothing to eat aboveground or when the weather is bad, these old males eat the seeds and the bulbs, while the other members of the colony have little or nothing to feed on.

Mature ground squirrels mate about ten days after coming out of hibernation. Three weeks later the young are born in litters that number as few as 2 or as many as 7. Their birthplace is another especially prepared room that the mother lines with soft grasses.

At birth a baby ground squirrel is naked and pink, blind and toothless, and weighs ¼ to ⅓ of an ounce. The development of this animal is so rapid that at the end of five weeks it leaves home. But full growth is not attained until the squirrel is two years old. By then it weighs 14 to 18 ounces, and measures 13 to 16½ inches from the tip of its nose to the tip of its bushy little tail.

The first home of a young ground squirrel is one of the numerous, abandoned burrows in the colony. The five-weeks-old animal moves in, and sets to work getting the home in order.

At the end of three months the squirrel has cleaned out some of the old tunnels, has dug a small hibernating den, and has lined it with grasses.

In addition to all this work the ground squirrel manages to eat enough to become fat enough so that there will be sufficient energy to sustain life during the first long sleep. As immature ground squirrels estivate and hibernate for longer periods than older animals, the first sleep may last thirty or more weeks.

EASTERN CHIPMUNK

As a rule animals that store food to eat during winter are not animals that hibernate. But almost every rule has at least an exception or two. And among animals the eastern chipmunk is one of the exceptions.

A bright-eyed little animal with a striped coat of rusty-red, the chipmunk is busy all summer long stocking its storage rooms with various foods. As acorns, nuts, and seeds ripen, the chipmunk gathers them.

The chipmunk stuffs these foods into the cheek pouches with its forepaws. It does not fill one pouch at a time, but loads by putting food into a pouch on the right side, for instance, and then into one on the left. When both are filled so that each holds a load of equal weight, the chipmunk scampers off to the special pantry rooms in its burrow.

Pantry rooms are not the only spots in whch food is stored. The chipmunk even places some under its bed of dried leaves and grasses. In fact a chipmunk is apt to store so much food in this spot that the bed is often ceiling-high by fall. But by spring the bed may be at floor-level, if the chipmunk has been particularly wakeful or unusually hungry.

For its bedroom the chipmunk selects the largest of the chambers in the burrow system. The bedroom, the storage rooms, and the one used only as a toilet are at the end of a tunnel.

Some chipmunks dig tunnels that are as long as 20 to 30 feet. These tunnels are probably the homes of animals four to five years of age; chipmunks this old are nearing the end of their lives. Other chipmunk tunnels are as short as 5 feet. These are most likely to be the work of young animals.

A tunnel opening is usually concealed. It may be inside a rotting stump, beneath a stone wall, or hidden among fallen leaves. The first section is vertical, but then at a depth of 5 to 6 inches it angles off in any direction. The diameter of the vertical shaft and that of the long part of the tunnel is about 2 inches. And the bedroom, lined with leaves and grasses for insulation, is about 1-foot square.

Before the chipmunk retires for good, it goes through a sort of prologue to hibernation. It does not frisk about with its usual high spirits. It stays near home and eats less and less, though food may still be plentiful. And as activity and appetite decrease, the desire for sleep increases.

When it can no longer stay awake, the chipmunk drags itself to the bedroom. This trip underground is the last of the season. Permanent retirement is usually in mid-September, or when the thermometer registers a temperature of 50° F. for several days at a time. This is the temperature at which most hibernators grow drowsy.

Once the chipmunk settles down in its bedroom, it assumes either of two sleeping positions. Sometimes the

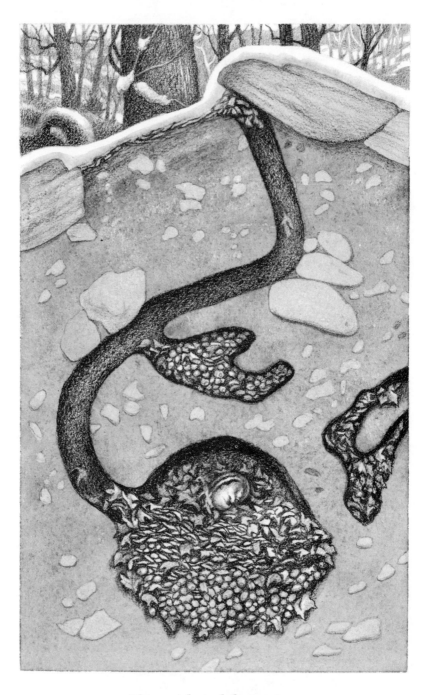

Chipmunk in hibernation.
Cross-section through burrow.

asleep-on-its-feet little animal arranges itself in a ball-like position. The head is brought forward and down and around so that the nose touches the belly. At other times the chipmunk sleeps curled up sideways like a kitten.

But no matter how it sleeps, the bright little eyes are always tightly closed. Then, as with other hibernating mammals, the life forces slow down little by little. In time they function at a level that barely sustains life.

Chipmunks live in hardwood forests throughout a great part of the eastern United States and southern Canada. On those areas of the range where winters are long, cold, and severe, this chipmunk sleeps more soundly than its western cousin, who pops out of its burrow even in zero weather.

Chipmunks on the southern part of the range are apt to be fitful sleepers. They are the ones that tend to nibble away at the stored food under the bed. Some come out of their burrows on warm days. And though it is contrary to the accepted winter behavior of more northerly chipmunks, a few have been seen aboveground in midseason in Massachusetts.

In a fresh fall of snow you can see the imprint of their tiny paws. The tracks they leave behind them look like those made by a small, resident bird such as the white-breasted nuthatch. The tracks show the impression of all five toes on each front foot.

For the better part of five months, however, most chipmunks are not among those present in any winter wildlife scene. But by the time March is here most of the snow has come and gone on much of the chipmunk's range. And by March, too, the sun and the rains have warmed the soil, so that the roots of ferns and other wild plants begin their work.

Spring, whose official arrival is March 20, is further indicated by the appearance of an occasional high-spirited chipmunk. At this season a chipmunk is charged with energy. Gone is the slow motion of the preceding fall. Now, every move is on-the-double.

The chipmunk races along stone walls; darts in and out of any and all openings; or then, seemingly out of breath, pauses to sit upright on a flat rock.

From time to time, as if to tell you how good it is to be alive in the spring, a recently awakened chipmunk utters the loud "Chip!" "Chip!" "Chip!" from which it gets a part of its name.

Soon after emerging from hibernation chipmunks mate. In April, when the spicebush is yellow with bloom and the new leaves of the wild violet are pale green, the young are born. Litters number 4 or 5.

A chipmunk grows up in a month. And from the day it is able to take care of itself until it hibernates for the first time, it leads a solitary life, except at mating season.

Once it leaves home for good a chipmunk is so busy that it works from sunup to sundown. It has to get ready for the coming of winter. Only the hottest weather halts activity. At such times the chipmunk gets relief from the heat by going underground. But, weather permitting, it works ceaselessly until autumn. As it works it is always on the alert for various bird and mammal predators.

Autumn begins in the northern temperate zone about September 22. And though crickets still chirp beneath the Harvest Moon and wild asters still bloom along roadsides, the nights are often cool.

By now the work of the chipmunk is finished: the burrow is ready for winter; the bedroom is lined; and the food is stored. As soon as the temperature of the air cools to

50° F., and remains so for several days, the chipmunk goes underground.

Then, safe from its enemies and the elements, the chipmunk sleeps away the succeeding days—days in which active animals have to endure any fall of snow, the fury of the wind, or the bite of frost on a dead-still winter night.

JUMPING MOUSE

Other deep sleepers among North American mammals are the jumping mice. These whiskered little creatures, with extra-long hind legs and long, tapering tails, are known as "Kangaroo Mice."

This nickname is apt, for these mice can make standing broad jumps of 4 to 7 feet. And if you happen to startle one as you walk near it, this mouse may cover 10 or 12 feet in one leap.

We have two kinds of jumping mice on this continent. One, the grassland jumping mouse, is thought to live nowhere else in the world. The tail of this mouse is tipped with black. Our other jumping mouse has a tail tipped with white. This is the woodland jumping mouse—an animal that has somewhat darker fur on the back than the grassland species.

Various forms of jumping mice inhabit the northern two-thirds of the continent. Some live as far south as North Carolina in the East, and in the West the range extends to central California and into central Arizona in the Southwest.

The grassland jumping mouse inhabits territories from sea level to mountain-high country. You will find this yellow-brown animal in bogs and meadows, in swales and

Jumping mouse.
This mouse can cover 10 or 12 feet in one leap.

swamps, and along fence rows. Other places to look for it are at the edge of shrubby forest borders or in sagebrush thickets growing on river flats.

The woodland jumping mouse lives in forested sections of low, hilly, or mountain country. The territory of this mouse is always near water. And the ground is usually covered with plants that grow in shady places.

As it searches for food the jumping mouse moves around on its territory in a zigzag fashion. It leaps first to the right, then to the left.

The leaps of 4 to 7 feet are possible because of a tail that measures 5½ inches. Almost twice the length of the body, the tail is the means whereby this mouse maintains equilibrium when it lands on all four feet. The tail acts as a counterbalance by steadying the animal as it lands.

If a jumping mouse loses a part of its tail, the animal somersaults when it attempts to land.

The jumping mouse eats a number of foods. The diet includes the seeds of various grasses, flowers of daisies and asters, and the fruits of many wild plants. It also feeds on beetles and butterflies, spiders and centipedes, and decaying meat and fish.

In late summer the jumping mouse crams itself with these foods whenever it eats. Such stuffing adds layers of fat over the back, belly, and groins.

By the time it is ready to hibernate the jumping mouse weighs about one-third more than usual. The usual weight is $2/5$ of an ounce to 16 ounces.

When it is not eating to build fat reserves, this mouse works on its winter home. This is a separate establishment from the one it lives in during summer.

Warm-weather homes are tiny round nests of grasses and leaves in which there is a small chamber. The size of the round little room is no more than 2 or 3 inches in any dimension.

Generally warm-weather homes are built on the surface of the ground. But they may be placed in a shallow hole, so that the upper half is aboveground. Or the mouse may build in a tree hollow, a stone wall, or under a fallen log.

Winter homes are always in the ground. The jumping mouse digs a burrow in a bank or on a hillside. Locations like these ensure proper drainage. Then, well below frost line, the mouse scoops out a den no bigger than a tennis ball. A lining of soft dried grasses and finely shredded leaves completes the wintering quarters.

By the time the nest is ready for use, the jumping mouse is so fat that its movements are slow and deliberate. And though the fat for hibernation is a blessing, it is also a curse. It adds so much weight that the animal is no longer agile.

The inability of the mouse to move quickly makes it easy prey for a number of meat-eating creatures. Some of these predators are snakes, hawks and owls, and foxes, skunks, and weasels.

While the jumping mouse works and eats, its winter coat grows in. You would not be able to tell the two species apart unless you saw the tail tips of each. The darker fur on the back of the woodland jumping mouse is now practically the same color as that of the grassland jumping mouse.

The fattest jumping mice are usually the first to hibernate. The mice on the northern end of the range go to sleep a month earlier than those in the south. The last to settle down for the winter are the females who had litters in late summer or early fall.

When the jumping mouse goes into its winter burrow for the last time it plugs the entrance with dirt. Then it waddles down a dark tunnel to its lined nest, and curls up into a tight little ball.

The tail is wound round and round the fat, densely furred body. Then all the life processes slow down bit by bit. Finally, the mouse is in the deep sleep of true hibernation. It breathes about once in every fifteen minutes.

The jumping mouse hibernates until several weeks after spring arrives in its particular area. Like all hibernating mammals, the mouse shivers from head to foot as it becomes active again. But once fully roused the animal goes aboveground. Not even cold or rainy weather stops it.

Soon after waking in the spring the jumping mouse mates. During the last weeks in May, or in June, or during the first two weeks in July, the young are born. They de-

Mammals 37

During hibernation a jumping mouse may lose 30 to 35 percent of its weight (⅖ to 1 ounce).

velop so rapidly that at the end of six weeks they are fully grown. They scatter to establish homes of their own.

The first six weeks of its life and again at mating time are the only intervals during which this mammal has any companionship. Otherwise it keeps to itself on a range of about one to nine acres.

During summer it bounds around on this range with leaps that make it the champion broad jumper in its size class among mammals. And during winter, snug in its tiny den beneath the earth's surface, the jumping mouse is unaware of the weather on its range.

BEAR

The bear, known as "Smokey," that looks down from the poster of the United States Forest Service and says, "Only YOU Can Prevent Forest Fires," is a black bear. This may seem a little surprising, for Smokey has a reddish-

Black bear in winter den.
Snowshoe rabbit outside looking in.

brown coat. But the black bears of North America have coats of many colors.

These color phases, to use the scientific term, include white, whitish-blue, creamy-white, chocolate-brown, and also cinnamon or reddish-brown. And then there is the most common color phase of all, the black. This is the one most of us know and the one so frequently seen in many National Parks, Forests, and Wildlife Refuges.

This well-known bear has a jet-black coat of long soft fur. The chest usually has a good-sized patch of white and the snout appears as if it were covered with brown silk plush. And the short claws are rounded, whereas those of the grizzly bear are long and only slightly curved.

The smallest of all our bears, the black is also the most common and widely distributed. This species is not so numerous as it used to be, but in one form or another you still find the black bear in many woods and forests throughout the continent. Its range extends from the northern limit of trees all the way south to a region of north-central Mexico.

Other North American bears, the grizzly, certain polar bears, and the big brown bear of Alaska, as well as the smaller black or cinnamon bear, sleep through a part or all of the winter. And they are the continent's only large mammals to do so.

The sleep of bears is not comparable to the deathlike state in which jumping mice, ground squirrels, and woodchucks pass the winter. The breathing, heartbeat, and body temperature of a sleeping bear remain unchanged or vary but little from those recorded for the months when it is active.

The rate at which a bear breathes in summer is 4 or 5 times a minute. In winter the number of breaths taken by a sleeping bear is the same. And there is no marked difference in the heartbeat of an active or an inactive animal.

But there is some variation between the body temperature of a wide-awake bear and that of one asleep in winter. The summer temperature is 98° to 102° F. In winter it is 88° to 96° F. This is still high enough to melt any snow that drifts into the winter hideaway—cave, windfall, or swampy thicket.

Long before winter is due the black bear prepares for a season that is apt to be one of great hardship for all animals. As soon as wild foods are plentiful the bear eats huge amounts of anything that is available.

Black bears in New York have been known to eat the

fruits, leaves, or seeds of thirty-eight different plants. They also eat insects, the nectar and honey of wild bees, and many small mammals and on occasion even such large ones as deer.

Those bears living in and around National Parks and Forests get additional food. They receive many a free meal from the millions of visitors to these areas.

Gorging on wild foods, freeloading at the expense of tourists, eating refuse or practically anything else you can think of, help the animal to acquire a layer of fat all over the body. Sometimes there is so much fat that it is four inches thick.

As the bear fattens the fur becomes denser than usual and as glossy as new patent leather. By late fall or early winter the coat is at its best, and wraps the 300 to 400 pounds of an adult bear in a pelt that any hunter might like to have as a trophy.

In accordance with State regulations the black bear is hunted throughout its range during certain designated seasons. Hunters have supplied us with some information regarding the condition of bears at the beginning of winter.

When they dressed fat bears killed in the late fall or early winter, hunters noted that the stomachs were empty and shrunken into tight, hard knots. Apparently when a bear becomes just so fat, the stomach contracts and prevents any further intake of food.

Black bears in Minnesota and the big brown bears of Kodiak Island in the Gulf of Alaska are known to eat certain foods that form an obstruction in the intestines. This obstruction is called the "tappen."

According to guides on Kodiak Island the big brown bears here eat quantities of wild cranberries just before settling down in their wintering spots. The berries act as a laxative of sufficient strength to clear the intestines. Then these big brown bears, the largest land-dwelling carnivores in the world, eat tough roots. These form a plug, the tappen, that remains in the system until spring.

Tappens found in the stomachs of Minnesota bears were composed of pine needles and hair apparently licked from the animals' own coats. Many hunters believe that the size of the tappen determines how soundly and how long a bear sleeps.

Whether the tappen influences a bear's sleep is not actually known. But it is common knowledge that thin bears do not sleep so soundly as fat ones. Those with little or no fat sleep fitfully and often leave their wintering spots to prowl around in search for food. And, as a hunter in central New York discovered, even a bear deep in winter sleep can be roused.

This particular bear was a female with two cubs. They were asleep in a shallow depression and were blanketed by some leaves and a light fall of snow. The hunter had on snowshoes and, as he strode along, he put one foot down on the female. The bear woofed and heaved herself up on all fours. Then, as the excited hunter accidentally fired his rifle, she lumbered off without her cubs. They were picked up and turned over to conservation officials for study.

Among polar bears only pregnant females drowse away the weeks during which the weather is at its worst. Some find a protected spot in the tumbled piles of shore ice; others seek the shelter of a drift of hard-packed snow.

Among black bears the first to den are the females. If the weather is mild and food is plentiful, a female black bear does not settle down until midwinter—sometimes not until just before it is time to bear her cubs. The last bears to become inactive are those in the South.

Bears apparently need rest, even when there is plenty of food available and living conditions are in their favor. And there seems to be some basis for believing complete rest for these animals is a necessity.

A New Hampshire farmer, whose avocation was natural history, once had a pair of pet black bears. They were well fed and well housed. They lived in a weatherproof barn and their "den" was a stall partially filled with straw.

Though these bears were sure of food and warmth during the winter, their behavior in the fall was similar to that of wild bears. The two barn bears burrowed into the straw, but did not go deep enough to cover their ears. They dropped their heads down and forward, then wrapped their paws around their heads:

If the owner spoke to his pets, they stuck their heads out of the straw. For less time than it takes to write about the bears' reaction to being spoken to, they seemed interested. Then they yawned. By the time each yawn was completed, they were again back under the straw.

As before the tips of four black ears were left uncovered. Like miniature triangular markers the tips indicated that bears were sleeping there.

Frequently the spot in which a wild black bear spends the winter is on the north or the northwest side of a mountain. Other bears curl up under tree roots or crawl beneath wind-felled trees, and some den in hollows of standing trees. Thus, cubs may be born almost anywhere—provided there is seclusion.

One, 2, and less frequently 3 or 4 black bear cubs are born in each litter late in January or early in February. Among black bears in Wisconsin one out of each 200 litters contains six cubs.

A female delivers, her young while she is half-asleep

Black bear emerging from winter sleep in early spring.

The young are tiny when compared to their mother. No longer than 8 inches, they weigh no more than 8 to 10 ounces. Cubs are blind and almost naked at birth, and, except for the opossum, are smaller in proportion to their mother's size than any other mammal.

For the first two or three months of their lives, cubs do nothing but eat or sleep. All during this period the female eats nothing; there are many times when she could not get out of her den even if she wanted to look for something to eat. Snow and ice often seal her in until only a breathing hole is left open.

On some parts of the black bear's range, a mother and cubs emerge at about the time when the marsh marigolds are in bloom, but while the buds of the bloodroot are still wrapped in their gray-green leaves.

Although a female black bear takes no food for several months and in addition nurses her young, she has not lost much weight by spring. Males also come out of winter sleep with plenty of fat on their bodies.

Recently awakened bears move slowly and appear dazed. Perhaps the reason for the slow movements is due to the fact that their feet are tender after sleeping so long. They usually go to a stream and drink as if their thirst could never be quenched. But they never seem to be in any hurry to eat. For bears on the northern part of the range a lack of appetite may be a good thing. At this time of year food is scarce in the North.

As the season progresses and foods become available, bears resume full activity. Throughout the black bear's territory, solitary males and females with and without young amble about in search of something to eat. When night overtakes them, they sleep wherever they happen to be.

If food is plentiful during late summer and early fall, a mother and her cubs acquire the fat for wintering. Then the family settles down, spending a full winter together for the first and only time in their lives.

By June of the following year, at about the season when the fawns of the white-tail deer are being dropped, the female black bear is no longer interested in the eighteen-months-old cubs. She either deserts them or drives them away. From then on they lead almost solitary lives, except for a female with partially grown young.

BADGER

Some of our smaller carnivores seek safe retreats in which to drowse away prolonged cold or continued stormy weather. Among such flesh-eating mammals are the flat-headed badger, the bushy-tailed raccoon, and the well-known striped skunk.

Though these animals escape blizzards, icy winds, or below-freezing temperatures by sleeping for several days or a few weeks at a time, they do not suspend animation as do true hibernators. Their periods of inactivity are more like those of denning bears—the body functions remain normal or nearly so.

One of these on-and-off sleepers is the short-tailed badger. This silver-gray mammal has a flat head with vertical stripes of black and white. The claws on the fore-feet are much longer than the claws on the hind feet. The long claws on the front feet enable the badger to dig rapidly into the earth to escape danger, or to dig out prey with amazing speed. They are also used to tunnel its summer and winter homes.

Less than a century ago there were any number of badger burrows. They could be recognized by the telltale

Badger carrying snowshoe hare.

piles of earth at each entrance, measuring 8 to 12 inches in width.

Today burrows are much fewer and much farther between. Acre after acre of badger territory has been used by our civilization and so many of these animals have been killed as pests, that to see a wild one is unusual, if not rare.

Except for areas in several North Central States, the badger lives almost entirely west of the Mississippi River. In addition there are a few in south-central Canada. From this northern outpost the range extends south through some part or all of each state west of the Mississippi and into the northern states of Mexico.

How early in the season a badger begins its interludes of sleep and how long each one lasts depend upon the whereabouts of the animal on the range.

One living in northern prairie or plains country or in mountain territory at elevations of 3,000 to 5,000 feet goes below ground earlier and sleeps longer and later than those in less cold climates. In the central part of the range the badger does not den so early or so often and the intervals of sleep are only a matter of days. And toward the south the animal is active all year round.

By fall a northern badger is usually well-layered with fat. And by the time winter has set in, it is not so hungry or so wide awake as at other times of the year. Apparently there is a seasonal check of some sort on appetite and activity. At the proper time this check decreases the desire for food and increases the need for sleep.

When the wind howls among the peaks and crags of our western mountains or drives the snow in stinging blasts across plains and prairies, the northern badger escapes such weather by holing up in a large chamber below frost line.

Winter quarters are at the end of a tunnel 5 to 30 feet in length and at a depth of 2 to 6 feet below the ground's surface. Loose dirt from within the tunnel is pushed up the passageway to plug the entrance. This "door" keeps out the cold and the wind.

Apparently a badger is not fully sustained by its accumulated fat during the intervals in which it sleeps. Neither does it seem to suffer a complete loss of appetite. The instant it wakes up, it is hungry. Immediately the badger shuffles up the passageway of its burrow, pushes away the earthen door, and comes out.

Whether it is so cold that the branches are snapping and cracking or so warm that ice and snow are melting makes no difference to a badger with an empty stomach. Once aboveground it shambles off with its toed-in gait in search of something to eat.

Food may be the carcass of a winter-killed deer or other creature that did not survive the season. Or it may be such live prey as rabbits, skunks, and ground squirrels. If it finds a dead animal, the badger eats its fill. If it is successful in a hunt, the kill is dragged into the burrow.

Then, with food at its side and shelter provided, the badger eats and sleeps by turns. It stays below ground until the last shred of its most recent kill is consumed. If, at any time during the year, a badger kills more than enough to satisfy its hunger, it buries the excess to eat later on.

One summer a badger in Yellowstone National Park dug out a number of ground squirrels. Some were young; others were half-grown; and still others were adults. The park badger ate only the smallest ground squirrels, and, after killing the large ones buried them.

Of course no one knows whether this badger could recall

winters when food was scarce. It would seem as if it did recall sometime past when it went hungry.

RACCOON

A farmer in Missouri learned that the gray-brown raccoon is not at all particular about what it eats if the food helps to put on winter fat.

Each spring the farmer planted watermelon seed on his farm near Eagleville. And each fall he harvested a melon crop that was good, but not so good as he wished.

One day he read that certain Indians fertilized corn by burying a dead fish in each hill. He decided to try the Indian way of growing things on his watermelons the following season.

The time finally arrived when the weather was right to put in watermelon seed. First the farmer caught some bluegills, a species of sunfish that he had stocked in his farm fish pond. Then, as he seeded each hill, he buried one bluegill in each little mound of earth.

The farmer never learned the outcome of his experiment. Raccoons, like a band of masked robbers, staged a raid one night on his newly planted melon field. As fast as they could, the 'coons dug out and ate the bluegills from each and every hill.

Melon-patch fish and those in streams are not the only foods the raccoon likes. This stocky animal, with pointed ears and ringed tail, eats almost anything, anywhere. So it is fairly safe to say that the 'coon is all-eating, omnivorous, and also that it is all over the place, omnipresent.

Some kind of raccoon lives in all forty-eight states, a small area of southern Canada, and a region of northern Mexico. In most open woodlands on this extensive range, the 'coon climbs, walks, and swims as it searches for prey.

From early spring until late fall a northern raccoon eats a great variety of plant and animal foods that grow in forest, field, and stream. A large and varied diet is also typical of the southern species—animals that do not den in winter.

Each day the northern raccoon leaves its den in a hollow tree at about the time owls first begin to hoot. It feeds all night over a territory that is always within easy reach of water.

As the season progresses and foods become more plentiful, the 'coon eats more and more. Its fondness for either green or dried corn is seemingly never satisfied. And in areas where acorns are available a 'coon gobbles down these nuts as fast as it can.

The enormous day-to-day feeding and the gorging on such rich foods as acorns put on fat. The layer across the rump and the back is often one inch thick. It is covered by darker, denser fur than that of the summer coat.

When winds blow cold across dry, brown fields or whip through the bare or nearly bare branches of trees, the raccoon clambers up the home tree and into the hollow containing its den. Once inside it curls up with the tip of the black nose half-buried in the soft fur of the belly. If the den is a large one, several families often share it. Such family gatherings include the most recently born young.

Denning raccoons are not hibernators in the true sense of the word. Theirs is not a drugged sleep in which they are unable to move, to hear, or to waken quickly.

Other animals that come into the den on purpose or by mistake are routed. A united group of wide-awake and thoroughly angry raccoons eject the intruder with all the dispatch of bouncers rushing disorderly people out of a public place.

Raccoon leaving tree-den to look for food in winter.

A heavy snowfall or below-freezing temperatures in November cause a raccoon to den. A thaw in January brings it out to look for food, even though there may be twelve or more inches of snow. And in late winter, when temperatures are in the twenties, Connecticut raccoons often leave their dens.

By January, and certainly no later than February, most adult 'coons are up and about. This is the mating season on the northern part of the range. In the South raccoons mate in December.

Occasionally there is an early summer mating. When it is time to den, the young of an off-season mating only weigh about 3 pounds. There are no fats to sustain them during the denning period and the fur is not dense enough to protect them from the cold. As a rule late-season arrivals do not survive the winter. But this valuable animal is able to perpetuate itself. Most females in the North mate early in the season. The young are then strong enough to survive even the most severe winters of a first year.

Though the raccoon spoiled the fish-for-fertilizer experiment of the Missouri farmer, the animal is good to have around.

These animals are a source of food in some areas; they furnish a living for trappers in various regions; and they eat quantities of harmful insects on all parts of their enormous range.

Such is the raccoon's value that if natural dens are not available, boxes for their use should be put high up in trees in areas where trees with hollows are few or wanting. The boxes serve as daytime sleeping quarters when the animal is active and provide the necessary shelters for denning in winter.

SKUNK

At about the hour drivers switch on their car lights the striped skunk is wide awake and out of its underground den. This brownish-black animal, feathery tail held high, is ready to start on an all-night ramble. In a thorough search for something to eat, it prowls around on a home territory of about 30 acres.

A striped skunk feeds on small rodents such as mice, insects of many kinds, and corn-on-the-cob, as well as a number of other plants and animals.

If there is plenty to eat early in the season, a skunk is acquiring fat by midsummer. And if there is no food shortage in the months to come, this largest member of the skunk family is as fat as a well-fed pig by fall.

Then, when the weather is no longer to its liking, a skunk seeks shelter in a dry, frost-free den. As the range of this animal includes the greater part of North America, the time of denning varies on the northern part of the range. Southern skunks are active all year round.

Denning is influenced by the temperatures prevailing in the skunk's neighborhood, and also by sex and age.

As soon as the thermometer registers 50° F. or less for several days in succession a young northern skunk of either sex becomes drowsy and goes underground for most of the winter. As the cold increases the next to retreat is the adult female. The last to quit the outside world is an adult male. But it takes earmuff-and-mitten weather to drive him below ground.

For some skunks the winter sleep lasts only a couple of weeks; for others the average is six weeks; and for still others it may be as long as twelve weeks.

Striped skunk about to enter burrow where others are already asleep.

So fat that he waddles and so sleepy that he totters, a full-grown male drags himself into his tunnel and makes a slow descent along it. A tunnel slopes gently downward until it reaches a depth of 6 to 12 feet, and ends in a small room below frost line.

A lining of dried leaves and grasses insulates the room. A bushel of insulating material is sometimes pushed down the tunnel and into a room by a skunk during the summer, when the animal is getting ready for the cold months ahead.

So much insulation is used on occasion that in Iowa a skunk chewed off all the grass for a radius of 8 feet around the entrance to its tunnel. As a rule entrances are not in the open. They are hidden by some sort of cover.

The room in which a skunk passes the winter is usually at a greater depth than the one used in summer. A skunk may dig its own tunnel and room or use those abandoned by a badger, fox, or woodchuck.

Some skunks winter under farm buildings or even beneath heaped-up stumps or piles of rocks culled from a field. Perhaps such animals are atypical in that they do not follow the usual winter behavior of most skunks.

Other skunks are solitary sleepers and the sole occupants of the room they dig and line. But more seem to believe that there is safety in numbers and den in groups. These animals must sense that by sleeping together they conserve body heat.

Skunks in the same den frequently include one male and several females. But this is not the only combination. There may be about the same number of young and old; or there may be fewer young than old; or then, again, there may be more young than old.

Any combination of sleeping skunks does not remain constant during the winter. For skunks leave one den on

warm days to go outside. Then, when the weather changes, they seek another den in which to hole up.

Denning skunks make an easy winter meal for a hunting badger. Though skunks are roused by a badger digging swiftly into their winter chamber, they are not particularly alert. The badger kills any or all in the den with no trouble. It eats its fill and buries any that are left over for future use.

Mild weather lures male skunks aboveground as the promise of spring tempts migratory birds into starting their flights back North. Such weather even brings out skunks in Maine and those in southern Canada. And only below-zero temperatures send them back to their dens.

Not long after Ground Hog Day, February 2, or earlier in the South, males are ready to mate. Extreme cold now or in March does not keep them at home. The search for a mate often takes a courting male as much as 5 miles in one night. Some males find mates aboveground, but others have to go right into winter quarters for theirs.

As skunk fur is at its best during February and March, the animal is trapped at this season in accordance with various state regulations. Thus the time of mating for skunks is the time of taking for trappers. A great many males and some females, whose winter sleep was brief, are caught. But safe from the trapper are the longer-sleeping females. Due to the winter behavior of these animals a new generation of skunks is born each spring.

It is our good fortune that some female skunks sleep longer than others. For skunks are one of our allies among animals. They eat quantities of mice, including meadow mice—great destroyers of farm crops. New York growers of hop vines, whose fruits are used for flavoring and in medicines, insisted that skunks be protected. Their reason was because skunks devour quantities of hop grubs that, unchecked, spoil the hop crop.

Probably a skunk is best-known for the musk it discharges to protect itself. Some people are so allergic to the musk that they become violently ill. Others are little affected by it, and at a distance the pungent odor brings to mind the fumes of ammonia.

Whether you find skunk musk pleasing or distasteful, it, like the animal from which it comes, benefits us. Skunk musk or its artificial equivalent, *butyl mercaptan*, is used in Western mines to warn of fire.

When fire is discovered a few drops either of the real or artificial musk are dropped into the system that circulates air throughout the shafts and tunnels of a mine. The liquid vaporizes instantly and is then borne along on air forced through the ventilation lines.

These lines go to every part of a mine, the levels of which are hundreds or even thousands of feet deep below the earth's surface as are those of the Homestake mine. The lower levels of this South Dakota gold mine are at a depth of more than 6,000 feet. Once a miner on any level gets a whiff of the vaporized skunk musk or its artificial equivalent, he knows that fire is to be reckoned with.

The skunk occasionally eats the eggs of domestic and wild fowl and some birds themselves. It spoils some farm and garden crops and devours bees that make various crops possible through pollination. But the good it does more than offsets the harm.

One outstanding service performed by the skunk is the destruction of snapping-turtle eggs. The savage snapper preys on ducklings, and each year kills a surprising number as they paddle around on the water near which they were hatched.

If it were not for the skunk, a great many more would be killed. The skunk often digs the eggs of this turtle out of their sandy nests and eats all of them before the sun's warmth can hatch them.

In addition to this fine conservation practice the skunk is one of the best catchers of mice and rats that you can have around. The animal, whose scientific name means "noxious odor," is therefore a good neighbor, due to its good work, good disposition, and good social habits.

Birds

BIRDS HAVE been called "glorified reptiles." There is good reason for this designation. Fossil remains of birds with characteristics peculiar to both animals have been found in southern Germany.

These fossils show tails, jaws with teeth, and clawed wings. But even with such reptilian features, the feathers on these fossil birds were well developed. Two of the most notable ways in which birds still resemble reptiles are the scales on the lower parts of the legs, or shanks, and the habit of laying eggs.

The encyclopedia defines a bird as a "warm-blooded, egg-laying vertebrate animal, having its body covered with feathers and its forelimbs modified into wings."

A bird's feathers correspond to the scales of a reptile, and, like scales, feathers are arranged in definite patterns. Though they are light in weight, feathers nevertheless protect against wet and cold. The hind legs are used as we use ours: to climb, to swim, or to walk. Usually each leg has four digits, one of which extends backward. And the skeleton of a bird is strong and light, with air spaces in many of the bones.

The heart of a bird is similar to that of a mammal. In both animals this organ is four-chambered. But the body temperatures of the two animals differ. That of a bird is 2° to 14° F. higher than that of a mammal.

There are 8,500 to 9,000 species of birds in the world. Some 700 to 800 are in North America, north of the Rio Grande. These birds live in all sorts of habitats, from the tundra in the far North along Bering Sea; south through mountain and plains country, inland and coastal marshes, and woods and forests of all types; to the subtropical regions of the Southeast and the deserts of the West and Southwest.

Many of these birds, such as the ruby-throated hummingbird, spend a great deal of time in the air. Some, such as the loon, pass the better part of their lives on or in the water. Still others, such as the pheasant or the quail, are more at home on the ground than any other place. And with but one exception on record, none of these birds hibernate, though some undergo startling temperature variations within twenty-four hours.

POOR-WILL

The name given to the poor-will by the Hopi Indians is more appropriate than they or anyone else ever dreamed. The Hopis call this grayish-brown bird "Holchko," the sleeping one. And much to the astonishment of everyone who thought *all* poor-wills migrated, there are a few of these birds that justify the Indian name.

Until recently none of the scientists who make bird study, ornithology, a profession believed that any bird hibernated. But after World War II, an ornithologist in

Poor-will.

California reported that he had discovered a bird that appeared to be hibernating. The discovery occurred in the Chuckwalla Mountains.

These mountains are a part of the summer range of the white-throated poor-will. Nature has gouged out deep canyons in the Chuckwallas and veined them with many crevices. On a December visit to the mountains a scientist discovered a sleeping poor-will.

The scientist could hardly believe his eyes. This was no time of year for the migratory poor-will to be around. It should have flown off long ago with other migrants to areas where there were plenty of insects. But what this poor-will should have done and what it was doing were two different things.

The poor-will had fitted its 7½ inches into a rocky opening. First, the scientist held a cold mirror in front of the poor-will's nostrils. No moisture collected on the mirror. Then he beamed a strong light into the pupils of the bird's eyes. The poor-will did not even try to close the lids. An attempt to detect the heartbeat with a medical stethoscope indicated no heart action. Lastly, the scientist took the bird's temperature. The body temperature was more than 40° F. below the temperature for its active state.

These tests seem to prove that a bird was actually hibernating!

A band to identify the poor-will was placed on its leg. Each year for the next four years the poor-will was discovered hibernating in its rocky niche.

But when the spot was visited in the fifth year no poor-will was found! Presumably it had been killed by a predator or had died a natural death elsewhere on its range.

Since the discovery of one hibernating poor-will, there have been two other reports of bird hibernators of the same species. It would seem, then, that the Hopi Indians knew something of the behavior of the poor-will when they named it "the sleeping one," though they were not aware that an occasional bird did hibernate.

HUMMINGBIRD

Among birds only the poor-will is on record as a hibernator. But another American bird undergoes changes when resting that are similar to those which occur when an animal hibernates. This is the hummingbird.

In a family that is closely related to that of the poor-will, the hummingbird is one of the smallest birds of the Western

At night the ruby-throated hummingbird becomes cold, immobile, and unable to fly.

Hemisphere. It is also one of the most active. And because it is so active the hummingbird feeds constantly, flitting back and forth among various flowers to sip the nectar.

The hummingbird can hover in a horizontal position by beating its wings 55 times a second. If it increases the wingbeat to 75 a second, it can fly forward at about 60 miles an hour. Hovering and flying 60 miles an hour consume great amounts of energy.

When a hummingbird is actively hovering, the energy it uses to maintain this position is about six times greater than when it rests. And during a daytime rest the amount of energy consumed is more than that recorded for any other animal.

But when the hummingbird rests at night, the energy consumed is about one-twelfth of the amount required during a quiet daytime interval. At night the little bird becomes cold, immobile, and unable to fly. It seems to be in a state similar to hibernation.

Reptiles

REPTILES BELONG to a class in the animal kingdom between the birds and the amphibians. There was a time when reptiles were the most numerous form of animal life. They were creatures of the air, the sea, and the land. Today, there are only a few remaining species of reptiles. All behave in the same way toward their young. As soon as the offspring are born the parents abandon them.

A snake is a legless reptile whose body is covered by smooth scales or scales that are ridged. Though all scales are a body covering, some also serve another purpose. By means of the action of the larger belly scales, the snake glides over the ground at a speed of about one quarter of a mile an hour. This is as fast an any of our 130-odd snakes can travel.

GARTER SNAKE

The striped and spotted garter snake, like all other reptiles and amphibians, has no body temperature of its own. The activity of the animal is thus controlled by the temperature of the surrounding air.

When the mercury climbs high in a thermometer, the garter snake moves away from the heat. It slithers off to a cool spot, so that it will not die from exposure to the heat.

But when the mercury registers temperatures lower than 50° F., the garter snake becomes inactive. It then seeks a place in which to hibernate.

Of all our harmless snakes, this one is the most numerous, the most frequently seen, and of the widest distribution. Some form of this brown, green, olive, or blackish snake lives in every state and southern Canada.

The garter snake is the first snake to appear in the spring and the last to hibernate in the fall. At this time of year these snakes gather in great numbers at one spot. The gathering place is often the south side of a rocky hill. Such a spot may have been used by these snakes for many years.

You can tell the gathering place of snakes by the condition of the narrow cracks among the rocks. The edges of these cracks are smooth. This is because generations of snakes have pulled themselves to and fro by means of their belly scales.

The large scales on the greenish-white or yellow underside of the garter snake are attached by a small muscle to the free end of each rib. When the garter snake exerts rib pressure each scale rises. Then the snake wriggles forward with the gliding motion we all know.

If the weather is warm and windless, the snakes bask in the sun. But as the nights become cooler and cooler the animals find it less easy to uncoil. Their muscles are stiff. When the daily temperature is below 50° F., the snakes finally hibernate. They crawl into the cracks among the rocks or tunnel under them if the ground is soft.

Ready-made retreats or those dug by the animals must be well below frost line. If they do not extend down 3 or

Garter snakes congregating to hibernate.

more feet, the snakes freeze to death. In those parts of North America where the winter is mild, garter snakes do not gather in great numbers to pass the winter in hibernation dens.

While it hibernates, the garter snake does have wakeful intervals in which it searches for prey. It also emerges from the den during an annual breeding period early in the spring.

The male has tiny sensory organs on his chin. To court the female he rubs his chin along her back. Shortly after the courting and subsequent mating the snakes quit the hibernation den for good.

Some garter snakes leave as early as March. By the time the weather is really warm there are hardly any snakes left near the wintering spot. They have scattered to the summer feeding grounds. These areas may be as much as two miles from the spot in which they wintered.

The garter snake lives in various environments. You may find these lively little snakes in coastal marshes and inland swamps. You will also see them on farms and in mountain country. They are even in large city parks. And when I lived in the suburbs of Troy, New York, I found them in the grass of my own front yard.

Though it is a member of the boa family, the garter snake does not squeeze its prey to death. Neither does it kill by injecting poison in the manner of rattlesnakes. This snake shoots its head forward, seizes its prey, and holds it fast on the long rear teeth. Prey is swallowed whole and digested by strong, quick-acting gastric juices.

About the middle of August the female garter snake bears her young. Live birth among snakes or other animals is called "viviparous." Animals that hatch from eggs have an "oviparous" birth. And nearly three fourths of our snakes hatch from eggs.

The average number of garter snakes in a litter is 20 to 50. But litters of 70 to 75 have been reported. The number of live-born snakes is usually greater than the number that hatch from eggs.

As soon as the snakelets are born the female abandons them. Most garter snakes in any litter do not live even a few weeks. They are preyed upon by various natural enemies. These include the big black-and-yellow king snake and other snake-eating snakes.

The survivors in any litter feed on earthworms. This is

about their only food from the time they are born until the time comes to hibernate for the first winter.

During the first and second full summers of its life, the garter snake almost doubles in length. It continues to grow throughout the rest of its life, but growth is at a much slower rate. Some garter snakes measure 36 to 44 inches in length. Such individuals are exceptions. The usual length for this species is 30 inches. This measurement is attained when the garter snake is ten to eleven years old.

The life span of the garter snake is based on the known ages of captives. Wild snakes probably have shorter lives —even if the summer feeding grounds have an abundance of food and cool, shady retreats and even if the hibernating dens provide the proper temperatures for winter survival.

BOX TURTLE

Another of our hibernating reptiles is the box turtle. This dark-brown creature of damp or wooded areas has an odd way of protecting itself. The bright-yellow under shell is bisected by a horizontal hinge.

When danger threatens, the box turtle, or land tortoise, closes the movable halves on either side of the hinge. The halves fit tightly against the edge of the high, rounded upper shell. Known as the carapace, the upper shell together with the under shell, called the plastron, protects the box turtle as the plates of a tank provide safety for those inside.

The animal pulls in its head and forelegs and the tail and hind legs. Once these are completely withdrawn, the turtle snaps shut the two halves of the lower shell. Predators are then unable to get at the soft parts of the animal within.

The box turtle seems to fascinate dogs of all kinds. But hounds in particular apparently consider this turtle a sort of

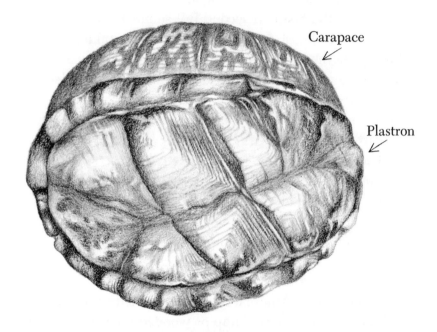

Box turtle seen from below, showing how lower shell (plastron) fits into upper shell (carapace) enclosing the body completely.

plaything. These dogs bark and sniff at turtles that have closed their shells. Hounds also worry shut-up turtles by flipping them back and forth.

A young hound in the District of Columbia carries home a box turtle in her mouth every time she is let out to run. There are often seven or more turtles lumbering around the house of her owner or in the yard and garden.

From time to time the hound's master gathers up these trophies of his pet's chase. He takes them out in the country, where he releases them. The freed animals shamble away on their bowed legs. They have to look for new territories on which there is food, water, and protection.

Let's follow one of these turtles, and let's assume that it was released in early summer. We'll trail a female, whose bright yellow eyes differentiate her from the male. His eyes are bright red as a rule, and his head is decorated by yellow blotches or speckles. These markings do not appear on the head of the female.

Our turtle takes a course that leads her into a field. Around this sunny, open spot are damp, woodsy areas. In the center there is a bed of wild strawberries.

The turtle finds the bed and pauses to nibble the small, bright-red berries. She selects the biggest and the best in each cluster, scissoring off tiny pieces between her jaws. This manner of eating is possible because the lower jaw fits closely inside the upper one.

Apparently strawberries are one of the box turtle's favorite foods, for our turtle will eat all that she can hold. She may eat so many that she will be unable to confine herself within her own shells.

If she tries to shut herself up, she just won't be able to do it. Her stumpy rear legs and her sprig of tail pop out between the carapace and the plastron at the rear if she manages to pull in her head and forelegs. And if she withdraws her hind legs and tail, then out in front pop her head and forelegs.

After gorging on strawberries our turtle rests. A spot where the ground is soft makes an ideal place for a nap. The turtle shovels into the soft earth with her hind legs and feet to loosen the soil. Then she twists and forces herself down into this especially prepared bed.

When our turtle wakes up, she starts off to look for more

food. This may include other wild fruits, various greens, including garden lettuce, and some insects.

In her unhurried travels this way and that she probably meets a male. The meeting usually results in a courtship. The male tags after the female as she wanders around, with a pause here and a pause there to nibble at whatever strikes her fancy.

This interval of slow-paced tag comes to an end when the male becomes somewhat aggressive. He gives the female gentle nips on the head and legs. If she ignores his caresses and continues to wander, he then uses force to stop her. He puts his forelegs on her back. Now she can't get away.

The two separate as soon as they have mated. Our turtle continues to move around on her territory until the first indications of cold weather. Then she selects a wintering spot. She looks for a place where the soil is dry and loose, or for one at the base of a tree where the leaf mold is inches deep. At either location she digs down until she is deep enough to escape frost.

In the vicinity of the District of Columbia, where our turtle hibernates, frost line is 10 to 15 inches below the surface of the ground. A turtle in Maine, Minnesota, or South Dakota has to dig much deeper. The records of the United States Weather Bureau indicate that frost in these three states often penetrates to a depth of 60 inches. In the Carolinas, frost seldom goes much deeper than 2 inches or 5 at the most. Digging in to hibernate for a Carolina turtle is a cinch when compared to the work a northern turtle has to accomplish in order to overwinter safely.

The longer the winter, the longer the period of hibernation for the box turtle. This is an interlude when the animal seems to be dead. Breathing is so infrequent that you might think that it had ceased. The chemical activity of the body also slows down so much that it, too, appears as if there were none. Even the development of the eggs that our turtle now carries within her body almost comes to a standstill.

Along about April our turtle digs herself out. Blossoming hepaticas, the flowers of the wild plum, and other signs of spring are evident as she ambles away from her wintering spot. The first trip of the year is to look for something to eat. The first meal and those for the next few days are enormous. Such feeding nourishes the eggs.

They now develop so rapidly that soon after coming out of hibernation the female is ready to release them. A shallow hole scooped out in soft earth makes a good "nest." The female expels 4 to 8 eggs. White and oval, they are somewhat elastic. Each one is covered with loose soil as soon as it drops into the nest. After all of them have been expelled the female pushes more earth over them.

The shallow hole in which the eggs are deposited acts as an incubator. Late in September or in October the eggs hatch. A young box turtle is no larger than a quarter. It wriggles out of the nest through the loose soil to the surface.

The yolk of the egg from which a turtle hatches is attached to the under shell. The newly born turtle feeds on the yolk for the first few days of its life. When all of the yolk is eaten, the quarter-sized creature looks around for food near its birthplace.

The first food of a young turtle is probably earthworms. They may be its only food during the first year of life. To get a drink it wades into the water until it is almost covered. Then it shoots out its neck, opens its mouth, and lets the water flow in.

Before the time comes to hibernate some turtles from

each yearly litter are killed. Skunks and other flesh-eating animals are the predators. The box turtle protects itself by hiding during the first five or six years of its life. Its shell is not thoroughly hard then.

But by the time the box turtle is ready to hibernate for the sixth winter, the shell is so hard that predators cannot harm it. If the box turtle is not killed by an automobile as it slowly makes its way across roads and highways, the animal may hibernate each winter for 20, 30, or 40 years. And it is a matter of record that a New England box turtle hibernated more than 130 times.

SNAPPING TURTLE

One of our best-known reptiles is the snapping turtle. This vicious, water-loving turtle belongs to the same order of animals as the gentle, land-loving box turtle.

Though both are of the same order and react to winter in the same way, the two turtles are quite different. The snapping turtle is no more like the box turtle than the poisonous water moccasin is like the harmless garter snake.

The snapping turtle has webbed feet and beaked jaws. It lives in still or slow-moving waters and in bogs and marshes throughout a great part of the continent. You can find these large turtles east of the Rocky Mountains, anywhere from southern Canada and Nova Scotia to the Gulf of Mexico.

When cold fronts from the North warn of winter weather to come the snapper hibernates. It buries itself deep in the mud of lake bottom or other body of water, or digs far down into the ooze of bog or marsh.

Encased in mud, the snapper exists in a state of suspended animation. The animal is barely alive. Its condition is like that of a person who has nearly drowned. By being buried alive in an almost lifeless condition the snapper escapes a season that it could not survive if it remained active.

An unusually warm spell may fool the snapper into coming out of hibernation too early. Then, when there is a drop in the temperature, the turtle is numbed by the cold. It cannot return to its hibernating spot. As a result it freezes to death.

As spring travels north months later, the snapping turtle rouses from its inactive state. And by the time early summer is here, the female is ready to lay her eggs.

The female comes out on land to wander around until she locates a spot where the earth is soft. First, she scoops out a shallow hole. Next, she lowers herself into the hole. Then, she wriggles around until the earth falls over her.

She starts to expel her round white eggs. She remains buried until the last of the 12 to 24 eggs are out of her body. Then she digs out at a sharp angle. A slantwise exit from the nest lets the earth fall back over the eggs. Now her work is done and the female returns to the water. She won't leave it again until egg-laying time the following year.

Snapping-turtle eggs are often discovered by skunks and raccoons when these animals dig for insects. Both animals eat all the eggs they can.

Unmolested clutches of snapping-turtle eggs hatch in eighty to ninety days. At birth a young snapper is black. By the time it is an adult this turtle has undergone a color change. The upper shell is a dull olive brown and the under shell is a dull yellow. Though the under shell of the snapper is similar in color to the under shell of the box turtle, it is not hinged.

To protect its dark head, strong legs, and tapering 11-inch tail, the snapper folds them back under the margin of the upper shell. On a good-sized adult the upper shell is 12 inches long and 10 inches wide. It projects over the 8-inch under shell; this affords some protection for the head, legs, and tail.

The meat-eating snapping turtle defends itself and also kills its prey in a much different manner than the box turtle. The snapper strikes like a rattlesnake. The head is thrust forward with a lightning-swift movement, and the prey is seized. Bone-crushing jaw muscles hold on to the prey.

Fishes are caught by the snapper as it lies on the bottom of a pond or other water. This type of prey is eaten at once. Prey caught on a stream bank or from beneath the surface of the water is dragged to the bottom. Before it eats, the snapper pulls a duckling, gosling, or other animal to pieces with its front feet and teeth.

There may still be another way in which the snapper differs from the box turtle. In the early 1950's there was a prolonged dry spell in Pennsylvania. Day by day and week by week the water behind the Greens Valley Dam in the central part of the state dropped lower and lower.

The snappers quit the little water remaining behind the dam. They burrowed into the cracked and caked mud around the edges of the puddle—all that was left of a once good-sized impoundment. If these Pennsylvania turtles dug in to escape the hot and dry period, they were estivating, passing a part of the summer in a torpid state.

There are other American animals that estivate. The leathery alligator of the southeastern United States becomes dormant during hot, dry spells. And the narrow-headed crocodile of southern Florida behaves in this manner, too.

The snapping turtle catches some prey from beneath the water.

These reptiles bury themselves in the mud and stay there until heavy rains release them.

If North America's alligator and crocodile estivate, then it is also possible that the snapping turtle behaves in the same manner. So when the weather is too hot or when the weather is too cold, the snapping turtle either estivates or hibernates. In these ways the animal escapes conditions that it could not survive if it remained active.

Amphibians

AMPHIBIANS ARE a small group of animals that are in the process of adapting themselves to life on land. The smooth skin of all these creatures is moist, and in most instances the skin is an important organ of breathing. Amphibians are the only group of backboned animals without some means of protecting themselves from predators.

TOAD

Frogs, newts, salamanders, and toads all hibernate. Some pass the winter in underground retreats. Others, like a tree salamander of California, overwinter in cavities of trees. One of America's most familiar hibernating amphibians is the squat brown toad.

The big-eyed toad, whose head, back, and legs are covered with warty lumps, is one amphibian that hibernates underground.

In the fall this toad digs a burrow. Sometimes the wintering spot is close to the foundation of a house. This location provides a little extra warmth to help the toad live through the winter.

Toad in hibernating position alongside the foundation of a house.

At about the time that the April rains are soaking the earth, the hoptoad awakes from its long winter nap. It digs its way to the surface. If the toad emerges in the spring of its third or fourth year, it starts out in the evening for the nearest pond or marsh.

The first arrivals at the pond are the males. They sit in shallow water, tilt back their heads, and fill their black throat pouches, or sacs, with air. Then they sing. The singing of the toads is a continuous performance for three or four weeks.

The males are soon joined by the females, who promptly

hop into the water. Although they are almost ready to lay their eggs, the females stage a sort of water ballet. They swim around and around to the accompaniment of the singing of the males. At the end of this performance each female lays several strings of eggs beneath the surface of the water.

Some strings contain as few as 5,000 eggs. Others may have as many as 15,000. Once the eggs have been expelled, the males fertilize them. Three to ten days later the eggs hatch. The length of time depends upon air and water temperatures.

The toads hatch as tiny black tadpoles. The *tailed* little creatures live in the shallow water at the pond's edge. Many are caught by birds, fishes, and other predators including turtles and diving beetles.

The tadpole stage lasts about nine weeks. During this period the future hoptoads are water-breathing animals. About the middle of July the tadpole stage ends. Through changes in form and structure the tail is lost, legs develop, and the gills are replaced by lungs.

Now the former water-breathing creature is one that breathes air. These changes in form are known as "metamorphosis." It is a characteristic of the class of animals to which the toad belongs.

The name of this class is "Amphibia," a term that comes from the Greek. "Amphi" means double and "bios" means life. And a double life is exactly what the hoptoad and other amphibians lead.

After the metamorphosis has taken place the toad stays close to the water's edge. It hides under bits of sticks, fallen leaves, or anything else that affords protection. Hiding is the only way in which the little creature can escape such predators as crows, skunks, and snakes. It ventures out to look for something to eat when the dew is heavy or if it has rained.

Once it has accustomed itself to land life, the toad hops off to find a territory of its own. The animal lives on its own small area for twenty or thirty years. And each year from April until fall the hoptoad feeds about four times during every twenty-four hours.

A toad usually begins to feed after the sun has gone down. It covers its territory in a hippety-hoppety fashion looking for such prey as May beetles, caterpillars, flies, and many other crop and garden pests.

The hunting toad moves in close to its prey, but stops a little short of it. The toad holds its head first on one side and then on the other. Apparently the animal lines up its intended victim as a man brings his target into focus between the sights of a gun.

The weapon of the toad is the long tongue. One end is attached to the front of the mouth. The other, the free end, is covered by a sticky substance. Once the toad is sure of its prey, the tongue whips out like a suddenly released jet of water from the nozzle of a hose.

Prey is hit by the sticky end. This holds the insect to the tongue as if it were glued there. The toad withdraws its tongue and swallows the prey.

The speed with which the toad flicks its tongue out and back is so rapid that you and I cannot see the action. Only a high-speed flash attachment on a camera can record it.

In a summer's feeding the hoptoad eats as many as 10,000 garden pests. The yearly value of the squat little amphibian is about $20.00 to a great many farmers and gardeners. For the toad lives almost anywhere on the continent, except in high mountain country.

The work it does in summer entitles the hoptoad to its

long winter nap. Apparently a lasting rest is necessary. The toad goes to sleep in winter—even when conditions are all in its favor.

Proof that the hoptoad needs more sleep than its usual daytime summer rest is shown by the behavior of some toads in a greenhouse. These toads were active during parts of the winter. But every so often they interrupted their usual routine by sleeping for days on end. This behavior seems to indicate that the hoptoad requires a certain amount of prolonged rest—even when conditions simulating summer are provided.

SPRING PEEPER

Early in April the noisiest and smallest frog in North America wriggles out from under old and rotting logs or from beneath the layers of leaves on the forest floor.

This clamorous frog is the one-inch peeper, called by scientists *Hyla crucifer*. The Latin name means "forest frog with cross on its back." This name is quite fitting; the spring peeper is a tree frog with the outline of a cross on its ash-gray or dark-brown back.

As soon as it emerges from its wintering spot, the spring peeper hops to a wet or watery place. Then it sings at the top of its voice. "Pe-ep, pe-ep, pe-ep, pe-ep," trills this tiny frog in shrill but sweet notes.

To sing the peeper draws air into its white throat. The throat swells and swells until it looks like a small blown-up balloon. Then, without opening its mouth, the peeper sings. After the last note dies away on the soft spring air, the peeper's throat looks like a deflated balloon.

You can hear the song of the peeper anywhere from Canada to Florida, and as far west as Arkansas. In fact you are more likely to hear one of these tiny frogs than you are

Spring peeper peeping.

to see one. They keep themselves well hidden. And in summer, when they are quiet, your chance of locating one is slim.

Summer is the time when the peeper conceals itself in low bushes, where it clings to stems and leaves. It is able to hang on because it has a rough pad on each finger and toe. From time to time the peeper comes down to the ground to hunt for tiny insects.

Like the hoptoad the spring peeper is born in water. It hatches from eggs laid underwater by the female. She fastens her eggs to the submerged parts of water plants, or, occasionally, leaves them unattached on the bottom, where they wash back and forth with the movement of the water.

Eggs hatch quickly, and for the next seven to twelve weeks the peeper is in the tadpole stage. By June or July at the latest the spring peeper has assumed its frog form. Then it comes out on land.

The first year of land life is short. In two or three months the days grow shorter and the nights become cooler. Soon it is time to hibernate. Then, when the daily temperature of the air is too low to keep the spring peeper alive, our littlest and our loudest frog hibernates. In one of the soft spots within its tiny range the spring peeper buries itself so deeply that no frost can reach it.

Then *Hyla crucifer* lies in a state approaching death. The shrill, sweet "Pe-ep, pe-ep, pe-ep, pe-ep" is stilled until the advent of another spring.

Fishes

THERE ARE more species of fishes than there are of any other class of vertebrates. The class of fishes, *Pisces*, is divided into three subclasses, with the vast majority of these animals in one subclass. Though there are differences among the three subclasses, all fishes have gills, fins, and dermal scales. And except for a few that estivate or come out on land from time to time, all live continuously in the water.

CARP

During winter some species of fresh-water fishes retire to the deep parts of lakes and ponds. Here the temperature remains more or less unchanging, and here the fishes pass the winter with little or no movement. And they eat little or nothing during this inactive period.

Other fishes, such as the silver or gold carp, sink to the bottom of their pond or lake. The carp squirms around in the mud until its one to six pounds is practically covered. Some carp pass the winter in a sluggish state, but others become truly dormant and remain so until spring.

Carp burrowing into mud at the bottom of a pond.

The carp is not a native North American fish. At one time it was found only in the streams and rivers of Asia. But in the early 1800's this fish was taken to Europe, and then in 1877 it was brought to the District of Columbia. A few years later some carp were released in rivers and streams draining into the Mississippi River and in waters in California.

Now carp hibernate in nearly every one of our lakes, ponds, and slow-moving streams, provided that these bodies of water have mud bottoms and are in cold-weather areas.

As it lies partly covered by mud, the extent to which the carp hibernates depends upon the number of times it breathes in each minute. And the rate of breathing depends upon the temperature of the water.

If the water temperature is 57° F., the rate at which the carp breathes is 57 to 75 times a minute. But when the temperature drops to 38° F., the fish inhales and exhales only 28 to 42 times a minute. And if the temperature of the water reaches the freezing point, 32° F., respiration ceases.

At the freezing point the carp appears dead. But actually it is only in a state of suspended animation, when all the body functions have temporarily ceased. Experiments have shown that a carp can remain in this condition twenty-nine days. In fact biologists have learned that the body temperature can go as low as 33.8° F. before the carp freezes to death.

The carp does not even die when it is encased in a block of ice. This is only true when the fish itself is not frozen through and through.

When the ice goes out of a lake and the water warms to a little more than 50° F., the carp revives. The coming-alive process is a slow one. Little by little the organs revive until finally the fish resumes its active life.

Early in the spring carp leave deep water and swim to the shallows. They spawn in such water from April until the end of June. During this period the fish thrash around in the water. If you are along a lake shore at this time of year, you can hear the splashes and plops as the carp break water and then drop back in.

A good-sized female between two and three years of age often produces more than 2,000,000 eggs during one season. As soon as the adults finish spawning, they return to deeper water.

The eggs hatch in a few days. Young carp are called "fry." Shortly after hatching, the fry start to feed on vegetation that grows on the bottom of the pond.

During the first summer a carp may grow 4 to 8 inches in length and may attain a weight of ¾ of a pound to 1½ pounds. By fall the fish usually has enough body fats to sustain it through the first winter, but only if there is sufficient oxygen in the water.

Carp and other wintering fishes occasionally meet disaster when the water lacks oxygen. This usually occurs where there is pollution or a quantity of rotting vegetation. Little or no oxygen in the water is sometimes the result of few inflowing streams. For streams emptying into a pond or lake usually bring in well-aerated water.

If these conditions are present there is often barely enough oxygen for the fishes to breathe. But if the ice is thick on such waters and also covered by deep snow, there is even less oxygen. And there are greater amounts of carbon dioxide. As a result many fishes die.

Perhaps you have walked along a lake shore early in the spring and noted many dead fish, floating belly up. The chances are that these fishes suffocated. Then, when the water was ice-free, they rose to the surface—reminders that fishes, like people, need the right kind of environment to survive.

MACKEREL

Some of our marine fishes undergo a sort of hibernation. Among the species whose behavior is different in winter than in summer is the fork-tailed mackerel. This green and silver fish spawns along our North Atlantic coast.

During the summer the mackerel feeds on plankton. This is composed of great masses of tiny plants and animals. They are near the surface, or upper strata, of the ocean. To feed on plankton the mackerel swims along with an open mouth.

By the end of summer, however, plant growth ceases and animal life dies down almost to the vanishing point. Now there is so little food that the mackerel leaves the upper strata of the ocean. It swims to deeper water and lies here almost without moving. Not until spring brings about a resurgence of the plankton does the mackerel rise again to the ocean's surface.

BASKING SHARK

The basking shark is another marine fish that becomes inactive during winter. This North Atlantic fish is one of our largest sharks. It often grows to a length of 40 feet and weighs as much as 3 or 4 tons.

In summer the basking shark cruises along at about two knots an hour. Like the fork-tailed mackerel, it swims near the ocean's surface. The animal strains tons of plankton from the sea water with its long whalebone gill rakers. Every so often this shark interrupts its swimming search for plankton; it takes time off to float on the ocean's surface,

Basking shark and mackerel on the ocean bottom.

basking in the sun. This behavior is the reason for the animal's name.

But with the coming of fall the plankton on which the basking shark feeds almost disappears. Two biologists have estimated that there is not enough plankton in each cubic foot of water to furnish the shark with sufficient energy to swim even at its slowest speed.

Basking sharks caught in winter have been without gill rakers. This seems to indicate that in the fall the basking shark sheds the gill rakers, and is thus prevented from wearing itself out in a needless search for food.

At this season basking sharks are rarely in the upper strata of the ocean. Scientists believe that the animal retires to the vicinity of the continental shelf. This is a plain on the ocean floor at a depth of less than 100 fathoms; it is usually of soft mud.

The shark lies on the bottom, motionless, breathing infrequently, and with the life processes functioning at a minimum. What sustains the animal at this time are the carbohydrates in the liver. During this idle period the animal slowly develops a new set of gill rakers.

When the suns of spring warm the ocean's surface and make it sparkle, the basking shark has new, fully developed gill rakers. By this time, too, the plankton is again abundant in the upper strata.

So once again the basking shark is seen in the North Atlantic, taking a sun bath as it lolls on the ocean's surface. Sometimes it floats along, backside up, with the dorsal fin plainly visible. At other times it lolls belly up, its lighter underside turned toward the sun. Then when it needs to eat the basking shark plows through the water with its mouth opened like a giant scoop to catch plankton.

Spiders

ALTHOUGH SPIDERS and insects are similar in appearance, each creature belongs to a different order in the classification of animals.

A spider can usually be identified by four pairs of jointed walking legs, whereas an insect generally has three pairs. Another difference between them is in the body. The body of a spider has only two divisions, while that of an insect is separated into three divisions.

The forward division of the spider's body is known as the cephalothorax, the head and thorax taken together. On this part of the body are the eight legs, the mouth parts and their poison glands, and the eyes.

A spider's eyes are located at the front of the head. They are not compound, like those of insects. They are simple, that is they have only a single transparent eye covering, the cornea, whereas the eye of an insect has several coverings.

The other division of the spider's body is the abdomen. This has no joints and it is generally rounded and short. At the rear of the abdomen are two or more pairs of silk-spinning organs, the spinnerets. These are used to weave cocoons for holding spider eggs. And with them the spider also makes a nest and constructs a web to catch prey.

Among the 2,500 species of spiders in North America are the crab spider, the jumping spider, the trap-door spider, the wolf spider, and those known as the orb weavers. These and other species pass the winter in all stages of life: as eggs, as young within the eggs, as juveniles, and as adults.

BLACK AND YELLOW GARDEN SPIDER

One of our most common spiders passes the winter as young within the egg. This is the one-inch black and yellow garden spider. Known, too, as the writing spider, this species is marked by bands or dots of bright yellow or orange. The forward division of the body is covered by silvery hairs, but the abdomen is hairless and as black as the front legs.

In the fall the female black and yellow garden spider weaves an egg sac. This receptacle is in the shape of a pear; in size it is about equal to the dimensions of a hickory nut. The sac is suspended from the branches of a shrub or it is fastened to the top of a weed stalk.

This spider lays all her eggs at once. There are usually 40 or 50. As each egg is expelled the female dusts it with a powdery substance. This dusting gives the egg a coating that looks like the bloom on a plum or a grape.

The eggs are enclosed in a silken cup at the center of the sac. The cup, in turn, is covered by a layer of flossy silk. And for additional protection the female weaves another layer of silk around both the cup and the floss. This outer covering is tightly woven and brown in color.

Shortly after the eggs are laid they hatch. The young

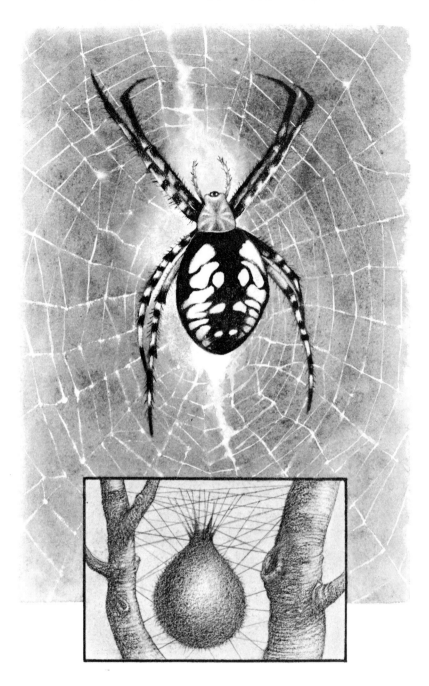

Black and yellow garden spider (female).
Length of body is 1″.
IN BOX: Cocoon within which the young pass the winter.

are known as spiderlings. They break out of the shells by means of an organ known as the "egg tooth." This later disappears.

Throughout the winter spiderlings are inclined to be dormant; they drowse away the days within their silken cup. During active periods they feed on what is left of the yolk of the egg from which they hatched. Spiderlings also feed on their weaker brothers and sisters.

By spring there may be only a dozen or so left out of the 40 or 50 that hatched the preceding fall. And even after the young spiders leave the egg sac, there may be a further reduction in their ranks as they are apt to prey upon one another.

A spider, like any other animal, has a territory of its own. To reach new locations the spider travels by a means of transportation known as "ballooning." A spiderling or spider throws out streams of silk. These threads form a sort of "flying carpet." It rises on warm currents of ascending air, and spiders and spiderlings are borne aloft and scattered far and wide.

Sometimes they go as high as 14,000 to 15,000 feet and travel hundreds or even thousands of miles. They drift to earth eventually and leave their flying carpets. You may have seen these threads caught on bushes and shrubs.

The discarded threads are known as gossamer or "Mary's silk." The name comes from an old European legend. Presumably, as Mary was taken to heaven after she died, her dress caught on some bushes. Several threads were pulled loose. The discarded silken threads of the ballooning spiders are supposedly the threads torn from the gown of the Mother of the Christ child.

The way in which a spider grows up is similar to the

stages by which an insect becomes an adult. A spider undergoes a series of moults before it reaches full size.

Some spiders moult four times; others undergo five; and still other species have as many as nine. As far as I know the exact number of moults that the black and yellow garden spider undergoes has not been determined.

To aid the spider in casting off its old skin, the animal has a liquid known as the moulting fluid. A spider forces it between the new tight-fitting skin and the old outer skin. The pressure of the liquid is an agent that separates the two skins.

Once the old skin is sufficiently loose, splits appear along the sides of the body and in front of the eyes. But no horizontal split occurs across the body. The vertical split along each side of the body and the one crosswise in front of the eyes form a flap of skin.

The spider pushes up the flap like a man thrusting up a hinged trap door. It pushes and pushes and pushes until the flap drops back over the abdomen. Out of this opening wriggles the spider. Each time a spider moults it gets all new hairs, and with each moult it grows larger. After the final moult the spider is fully grown. The sexes differ in size: the female is generally larger than the male.

The black and yellow garden spider is one of the orb-weaving spiders. All these spiders make elaborate webs, using their silk-producing organs and the spinnerets. Collectively the orb-weavers have seven different kinds of silk glands. Each is able to spin several kinds of silk. Not one of them, however, has all seven silk-producing glands. But every one has at least three.

The black and yellow garden spider is no lily of the field. It spins and spins and spins until it completes a gauzy, transparent sphere. A web is at times so large that the diameter may be 2 feet.

In late summer or early autumn you sometimes see the web of this spider strung in a bush or shrub near a house. More frequently you find them on marsh plants, meadow grasses, or on pasture plants such as the narrow-leaved sheep laurel.

Late in August or early in September the male leaves his web. He is seeking a female to court and searches until he finds the web of one. As the male does not want to be mistaken for prey, he drums and plucks on the female's web. If she responds to his drumming and plucking, they mate. Once the mating is over the female is apt to kill the male.

When a female kills a mate or prey she first bites it and then injects venom from the poison glands. The victim of the black and yellow garden spider is wrapped in a sheet of silk pulled from the spinnerets. When the female is ready to eat she sucks out the body juices.

Shortly after mating the female black and yellow garden spider is ready to spin the sac in which to lay her eggs. As we now know, the eggs of this spider hatch quickly and the spiderlings remain within the sac during winter.

The young, the half-grown, and the old of other spider species pass the winter in various places. You will find them under dead leaves on the forest floor, deep in beds of dry mosses, or even beneath stones.

As a rule adult spiders winter in special spots. They seek caves or crevices or crawl under the scales of bark on trees. These hibernating spiders often enclose themselves in a cocoon, or layer upon layer of fine silk. Their wrappings make thick blankets and are additional protection from the cold.

Most spiders are beneficial. The English naturalist John

Crompton has figured that the spiders of England and Wales alone eat twenty-two trillion harmful insects a year. And H. C. McCook, an American who made a lifelong study of spiders, says that man might not be able to survive at all without these creatures that overwinter in all stages of life.

Insects

As THERE are almost 700,000 known insects in the world, perhaps we should define what an insect is. And as there are more kinds of insects than all other animals put together, perhaps we should explain in what ways insects differ from other animals.

Insects are animals that in the adult stage have an outside or external skeleton. The body is also divided into three definite parts: the head, the part between the neck and the abdomen called the thorax, and the abdomen itself.

These animals have other characteristics that mark them as insects. Their eyes are usually compound, not simple like those of the spider. The body appendages include one pair of feelers or antennae, one or two pairs of wings, and three pairs of legs.

The growing stage of an insect's life is unlike that of other animals. All growth is made during the immature stages. But not all of it is made in the same way, for insects have two types of growth.

Some insects develop gradually. They undergo a series

of moults in which the skin is shed each time. After each moult the young more nearly resemble the adults. This form of growth is known as incomplete metamorphosis. A few insects that develop in this way are grasshoppers, crickets, plant lice, and various bugs such as the bedbug, the squash bug, and the stink bug.

Other insects undergo a complete change in becoming adults. The young or larvae in this category do not resemble the adults. When the larvae complete their growth, they enter a stage known as pupation. A change occurs now that is absolute. This is known as complete metamorphosis. Such insects as moths, butterflies, beetles, wasps, and flies undergo a complete metamorphosis.

Some of the insects discussed in *Winter-Sleeping Wildlife* undergo an incomplete metamorphosis. Others have a complete change. The insects selected for this book represent some of the more important orders. In addition they are insects that overwinter in different life stages and environments.

Most of them are species that you may see in your own back yard; on a walk across a field or through the woods; or as you wander along a brook or stream.

LADY BEETLE

The convergent lady beetle is one of more than 82,000 insects in the United States and Canada. This tiny insect is also one of the nearly 100 species of lady beetles in North America.

You can recognize this particular beetle by the black dots on the orange wing covers, an even number to a side and the basal dot near the head, and the two white dashes on the back. These dashes almost meet, or converge, hence the name of this insect. The quarter-inch creature is the friend of the farmer because it is the foe of many insects that damage crops.

Once hibernation is over, this beetle starts preying on farm and garden pests. During the growing season each of these polka-dotted beetles devours its share of mealy and green bugs, pink bollworms, thrips, and red spiders. Sometimes a single beetle eats 100 harmful insects a day.

There is a legend to the effect that the good work of the lady beetle was recognized as long ago as the Middle Ages. These valuable insects were dedicated to the Virgin and were called "the beetles of Our Lady." Today the more usual names for them are "ladybug," "lady-bird beetle," or "lady beetle." And today the value of these insects is still recognized.

Lady beetles are bought by agriculturists. The growers of various crops release the insects in orchards and melon patches, and also in fields of wheat, oats, and barley.

The beetles are sold by the gallon. It takes about 135,000 to fill a gallon container. Thirty thousand are required to keep in check the harmful insects on an acre of cropland. A gallon, therefore, controls pests on 4½ acres.

A California firm sells lady beetles to agriculturists, and the annual sales amount to more than 130,000,000 insects. They are shipped by air, and always at night, to all parts of the United States. Night shipments are an absolute necessity because the beetles are less active then.

Company employees collect the beetles while the insects are hibernating. The lady beetle of California has a different hibernation pattern than the species in the East. Each fall the California beetles fly to mountaintops for the cold season. They gather by the thousands in areas where there are shrubs and bushes and settle down on each and every twig.

A company crew goes mountain-high to collect beetles. It takes forty-eight hours for them to harvest a truckload. Back at the "beetle factory" the insects are put in cold storage. They can be stored for as long as two months at a temperature slightly above freezing. Soon after the beetles are brought out of storage for shipment they regain their usual liveliness.

In the East the lady beetle passes the winter in the tiny cracks of chimneys, foundations, or other weathertight sites. An ivy-covered wall is particularly favored. A spot like this usually has plenty of food available in the form of thrips. These are soft-bodied insects that suck the juices of ivy and other growing plants. They are also the bane of commercial flower growers who sometimes have entire plantings ruined by thrips.

Warm days in early spring bring out the lady beetle. Even warm winter days may cause the insect to interrupt its hibernation and venture forth. A late cold snap usually sends the little bug back to its crack or cranny to sleep for several more days—or even for several more weeks.

Sooner or later the lady beetle comes out for good, seeks a mate, then lays thousands of eggs. The larva that hatches has the appearance of a tiny carrot.

The body is covered by short, spiny tufts, and the legs are well developed. As soon as it hatches the larva starts to eat. The prey is other small insects, insects' eggs, and even small spiders including the red spider—another troublemaker for commercial flower growers.

The growth of the larva is rapid, and upon becoming mature, it assumes a more or less oval form. Now it is in its final stage, pupation, and hangs by the tail for a few days. After this interval of rest it emerges as an adult beetle, the *imago*. It is now in the form most of us recognize. And thus,

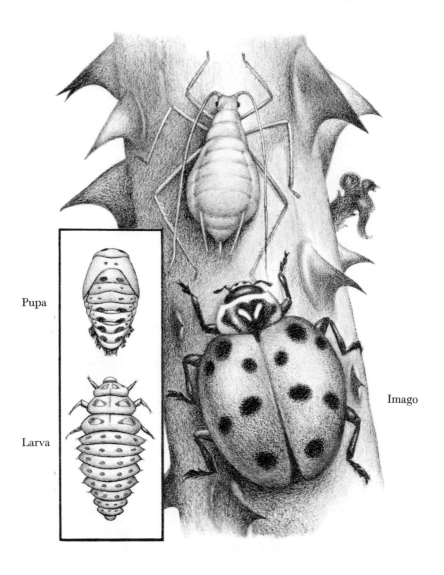

Lady beetle attacking an aphid on a rose stalk.
Actual size about ¼″ long.
IN BOX: *Bottom*—larva of lady beetle. *Top*—pupa of lady beetle.

aboveground, the lady beetle passes through all four stages of its life: egg, larva, pupa, and imago.

CABBAGE BUTTERFLY

Some of our butterflies do not migrate south to escape winter. The non-migratory species pass the winter months in four different ways. How a butterfly overwinters depends upon the species. It may pass the winter in any of the four distinct stages of its existence: as an egg, a caterpillar, a chrysalid, or as an adult.

One of North America's most common butterflies passes the winter as a caterpillar inside a cocoon. This is the cabbage butterfly, a little white butterfly that ranges over the greater part of the United States and Canada. During the caterpillar stage the insect is a pale green. Then it is often called the cabbage worm.

In the late fall this worm stops feeding on cabbage leaves. It wriggles off the underside of the last leaf in which it chewed hole after hole. It inches away to a safe place in which to pass the winter.

A stone, a fence post, a piece of board, or even the side of a building is suitable. To any one of these the cabbage worm attaches a cocoon.

To weave a cocoon the caterpillar uses an organ located in its head. This is the spinneret. With it and a substance from the silk glands, the caterpillar spreads threads of silk upon the surface of the wintering site. A loop is also fashioned at the same time.

As soon as these are completed, the caterpillar embarks upon its last activity of the season. The hind legs are entangled in the threads, and the body is inserted into the loop. Then, snug within the shelter of its own creation, the caterpillar is usually able to pass the winter in safety.

Cabbage white butterfly on cabbage leaf.
Actual size, wingspread 1¼" to 2".

Early in the spring the caterpillar bursts out of its pale green skin. Its body covering is now a gray or brown skin. For a week or two the animal seems lifeless. It is a chrysalid, a stage in its life when it is helpless because the skin enveloping it is firm. When the skin bursts, a butterfly is freed. Before it can take to the air the just-released creature has to dry and expand its wings.

The wings of this butterfly have a spread of about 1¼ inches. The forewings are black-tipped, and those of the female are dotted with black. A male can be differentiated from a female because his forewings have only a single black dot. Each sex has a single dot on the outer front margin of the hind legs.

Cabbage butterflies flit about over fields, gardens, and meadows. They pause to sip the nectar of early spring flowers. They feed by drawing the nectar through their long, coiled tongues.

From time to time the female alights upon a cabbage leaf or the leaf of other plants of the mustard family. During these pauses in flight she deposits her eggs. The eggs, tiny and yellow, are attached to leaves by a glue-like substance. During the two weeks of its life the female continues her egg-laying.

The eggs hatch in about seven days. The larvae are tiny caterpillars of a velvety-green appearance. They begin to feed at once. A recently hatched caterpillar eats for several days, then sheds its skin for the first time. Now it is larger and now it eats ravenously. During the next two or three weeks the skin is shed several more times. Finally the worm turns into a full-grown caterpillar. It continues to feed on the tender parts of cabbage leaves until the time comes to spin the cocoon in which to pass the winter.

WOOLLY BEAR (Isabella Tiger Moth)

Woolly bears, those fuzzy red-brown caterpillars that you see in the fall, are supposed to be weather prophets, and in some areas are still considered as such.

If the brown band on the woolly bear is wide, so runs an old, old superstition, this indicates that the coming winter will be mild. But if the band is narrow, then the winter is going to be a long one, with a great deal of ice, lots of snow, and below-freezing temperatures.

As a weather forecaster the woolly bear is no more reliable than the woodchuck. The width of the band is no clue as to the length and the severity of the forthcoming winter.

The woolly bear is the larval stage of the Isabella tiger

moth. Like any moth, it is a night-flying creature. As a moth it is not so well known to most of us as the woolly bear. You frequently see this caterpillar crossing roads and sidewalks on its range in northern states, from the Atlantic to the Pacific.

When the air is sharp with frost and the leaves cascade down, the woolly bear hibernates. It crawls under piles of leaves, trash, or other protective accumulations. In any of these spots the caterpillar curls into a tight little ball. It is not seen again until spring makes the days warm once more.

As soon as the woolly bear is out and about, it starts to eat. The first spring meal may be to the accompaniment of the cardinal's cheery song. As the notes of the cardinal rise and fall, the woolly bear nips off bits of leaves with strong biting jaws. It eats as much as it possibly can so that it will have body fats for nourishment while in the cocoon stage.

Suddenly the woolly bear stops all eating. It becomes inactive less abruptly. Little by little, however, it slows down until it finally lapses into a sort of sleepy state. There is a short, quiet interval before the animal becomes active again.

The woolly bear searches out a good spot for its cocoon. The underside of stones or discarded boards or occasionally old board fences are all used. Once a site has been selected, the woolly bear starts work on a cocoon. It uses the spinneret and the silk-producing glands to make a delicate, silky thread. The thread is woven around the body.

Wrapped in its cocoon of silk and caterpillar hairs, the woolly bear is now ready to change from a caterpillar into a moth—a transformation known as pupation.

Though there is no outward sign of what is taking place,

Woolly bear and Isabella tiger moth.
Actual size, wingspread 2″.

great changes occur within the cocoon. At the end of about two weeks the cocoon splits, releasing a moth.

The insect is in its final stage, the *imago*, or perfect and adult form. And instead of strong biting jaws, the mouth now has sucking parts.

The Isabella tiger moth has a wingspread of 2 inches. The body is a dirty orange with black spots. The forewings of the male are buff-brown and spattered with small black spots. The hind wings are straw-colored. The female usually looks exactly like the male, though now and again some have hind wings that are pink.

Shortly after the woolly bear becomes an adult, the female begins laying her eggs. Spherical in shape and yellow in color, the eggs are laid in patches on the leaves of plants. Before she is through, a female often lays as many as 1,000 eggs.

The eggs produce the crawling, grublike creatures that we know as woolly bears. The young are apt to have more black on the ends than the older ones. And the young are more inclined to stay together than those that are older.

The active life of the woolly bear is devoted solely to eating. The plants on which it feeds are of no economic importance. Its enormous feeding causes the woolly bear to grow rapidly. By the time it hibernates the creature is of giant proportions in comparison to its size at birth.

GRASSHOPPER

The number of destructive insects in the United States and Canada is nearly 10,000. The damage to crops caused by these insects in the United States and the Territories is $20,000,000,000 each year. And in 1955 the damage to Nebraska's farm and garden crops amounted to more than $2,400,000.

The insect that caused so much damage in Nebraska is the well-known grasshopper. With the exception of the butterfly, no other insect is so frequently seen in meadows and fields. A big-eyed creature, the grasshopper has long feelers and long hind legs that are thickened at the base.

This type of hind leg permits the grasshopper to outjump the kangaroo, if you relate the distance jumped by each animal to the weight of each. If ounce for ounce, the kangaroo could leap as far as the grasshopper, then the kangaroo would be able to bound along at the rate of a quarter of a mile. This would be a leap of 1,320 feet!

The insect that is so destructive and such a leaper usually passes the winter in the ground in the egg state. In the fall the female seeks a protected spot in which there is soft earth. She drills tiny holes that look as if they had been dug for miniature fence posts.

Grasshopper depositing eggs in the fall.
INSET: *Young, hatched the following spring.*

On the tip of the female's abdomen is a long, swordlike organ. This is the egg-laying device known as the ovipositor. The female uses it to insert her eggs into the holes she drills.

When the eggs hatch in the spring, the young grasshoppers bear little resemblance to the adults. A grasshopper undergoes several changes during its life cycle. But even with all the changes, this insect has only an incomplete metamorphosis.

A grasshopper sheds or moults its skin several times during its short life. Each time a skin is shed, the insect comes out in a larger skin. And each time this occurs, the animal more closely resembles the adult form. After its final moult the grasshopper emerges in the adult form.

A fully developed grasshopper has four wings. The hind pair are folded fanlike under the fairly long and narrow forewings. The antennae are long and so are the legs. Leaping tremendous distances is not the only use to which the grasshopper puts its long hind legs. A male also makes "music" with them.

A close inspection of the hind legs shows how the grasshopper produces the familiar crinkling notes. On the thigh are some pointed plates that overlap. And on each side of the shank are several sets of horny plates.

When a male courts a female, he plays music on his hind legs. He brings the shank up from the thigh, then rubs each leg section back and forth against the wing sheaths. He fiddles first with his right leg, then with his left. There are pauses of equal length between music from the right side or music from the left when the grasshopper switches his fiddling from one side to another.

The fiddling of the male ends with the first frosts. Both the males and the females die with the summer season. But safe in the ground are the eggs that will hatch next year's grasshoppers, once winter is over and once the warm days of spring signal that for the grasshopper life is possible.

PRAYING MANTIS

One of North America's most remarkable insects is the praying mantis. The name comes from the position the insect assumes while it lies in wait for prey. At such times this greenish insect raises its long, spiked forelegs in front of the shiny triangular head. So poised and perched motionless on twig or leaf, the animal appears as if it were in an attitude of prayer.

The Greeks called the insect "the divine." The Provençals in southeastern France refer to it as "the creature which prays to God." The Swedish botanist Carolus Linnaeus classified it as *Mantis religiosa*. And in everyday language we speak of it as the praying mantis.

The praying mantis is an insect that overwinters in the egg state. But the eggs, unlike those of the grasshopper, are not deposited in the ground.

In late summer or early fall the female lays clusters of eggs in the branches of trees and shrubs. The eggs are encased in a foamy mass; it hardens in a few moments after exposure to the air. This insulation prevents the eggs from freezing in the cold months to come.

The eggs hatch in the spring. The young are about the size of mosquitoes. In form they are miniatures of the parents, but in color they are black instead of green. And they are wingless.

As soon as it hatches, a young mantis starts eating plant lice. It grows rapidly and in a short time the color changes from black to green. By late summer the insect measures 3 to 4 inches in length.

Most of its life is spent perched on twig or leaf as it

Praying mantis attacking tiger swallowtail butterfly.

waits for insects on which to prey. A praying mantis devours many injurious insects. It eats some beneficial insects, too, including an occasional honeybee.

A section of the lower leg, the tibia, has a row of sharp spines. There are additional spines on the upper leg, the femur. When the praying mantis closes the lower leg knifelike against the upper leg, prey is held fast and cannot escape. This enables the predator to make a leisurely meal.

Any food particles left in the spines are cleaned out by the praying mantis. The insect also licks its tiny forefeet, then cat-fashion washes its rather solemn-looking face.

As it washes, the mantis turns its head first one way and then another. The ability to move the head is rare among insects. Only a few have a neck flexible enough to permit head movements of any kind.

Another physical characteristic makes the praying mantis unusual. It is the eyes. The insect has a pair of large, compound eyes and three smaller simple eyes. The small eyes are set in a triangle between the two large ones.

During the day the eyes are brown or green. But at night the pigment that gives them their daytime color is withdrawn. The eyes are then a lustrous black; this permits greater amounts of light to be absorbed.

By the time a female praying mantis is full grown it is usually larger than a male. Toward late summer or early fall the smaller male begins to court a female. Once the male has won a female, they mate. And once the mating is achieved, the female bites off the head of the male. There are times when the female eats the head of her mate. She never eats the body, however, but casts it away.

Now the female has only one mission left in life before she dies. She must lay her eggs before frost. The female often deposits her egg masses on a twig of the very bush on which she was hatched. At other times she goes to a different bush, or attaches them to the stems of grasses or brambles.

One female can deposit several egg masses. Each contains 50 to 500 eggs, and each mass is nearly 1 inch in diameter. Shortly after laying the last of her foamy egg masses, the female dies.

Then soothsayers, rear horses, or devil's horses, to use the other names for the praying mantis, are gone from the face of the earth. And not until the eggs hatch the following spring do we see "the creature which prays to God."

DRAGONFLY

If you have ever strolled along the bank of a stream in midsummer you have seen the big green dragonfly. This insect has great shimmering wings that can propel it through the air at a speed of 60 miles an hour.

As you watched the dragonfly wheel and dart, did you ever wonder what becomes of it in winter?

A young dragonfly is known as a nymph or naiad. And in its nymph form this insect overwinters underwater. It worms its way into the soft bottom mud of lake, pond, or slow-moving stream. Most nymphs winter only a few feet from shore, but some hide in the decaying litter or other debris farther out.

The young dragonfly nymph is sometimes active until the end of November. But by the time the ponds on the northern part of its range are sealed by a winter-long coat of ice, most nymphs are inactive.

The life process, the metabolism, that is the source of a nymph's energy is at low ebb. In fact the metabolic rate is so low in winter that the insect does not even moult.

During a nymph's active underwater life there are a number of moults.

When the suns of spring warm the water the nymph becomes active. The animal starts eating at once. It gorges on all sorts of underwater insects. Then some time in May the nymph stops eating altogether.

Motionless, it clings to the underwater part of a plant's stem. After several inactive days the nymph rouses. It starts to climb up the stem.

Inch by inch the dragonfly-to-be edges upward along the stem. Lighter and lighter becomes the surrounding water as the insect climbs higher and higher. Its head finally breaks water, causing a ring so tiny that it might have been made by the point of a pin. Then, bit by bit, the nymph continues its advance upward, until at last it clears the surface. Now, for the first time in the eleven months of its life, it is out of the water.

The still-wet nymph clings with its sharp claws to the stem of the reed. At first the creature is quiet. Then, like a mammal emerging from hibernation, it starts to tremble. Next a small slit appears below the head.

The slit gradually lengthens in two directions. One end travels forward until it reaches a point between the eyes. The other end spreads backward until it extends to a point between the wing pads.

This slit permits the adult dragonfly to emerge. The first part to appear is the back. Then out pops the big-eyed head. This is followed by the long legs and the crumpled wings. Last of all the needle-like abdomen is freed from the shell. The entire process is rather like the slow eruption of a volcano during which debris is tossed bit by bit from the crater.

At first the newly emerged dragonfly is not active. The

Dragonfly.
TOP: Adult which has emerged from the empty nymph shell immediately below it.
BOTTOM: A nymph crawling out of water to burst from its shell and become an adult.

insect hangs on to its last skin for an hour, perhaps two, or even a bit longer. Eventually the glistening colors appear; the wings dry and spread; and the insect gains sufficient strength to launch itself into the air.

The better part of the short life of an adult dragonfly is air-borne. All its prey is caught on the wing. And it has been well equipped to hunt in this manner.

The enormous eyes are highly developed. They are made up of as many as 20,000 sight units or facets. As a result the vision is excellent, and the dragonfly can see in all directions.

The muscles in the thorax contract and expand so rapidly that the wings move at 28 beats a second. And, finally, the six long, reddish legs let down and bend forward to form a sort of basket as the dragonfly flies along.

When a dragonfly snares a mosquito or midge, the prey is eaten at once. The head and body are quickly devoured, but the legs and wings are discarded. These parts fall away from the wheeling and darting dragonfly like chaff dropping from a threshing machine. So big an eater is the dragonfly that it often devours its weight in flies in half an hour.

As the dragonfly hunts, a number of animals in turn hunt it. At water level fish and frogs are ever on the alert to snatch a dragonfly as it skims low over the surface of pond, lake, or stream. And from the air kingbirds, swallows, and some small hawks are always ready to swoop down on a hunting dragonfly.

The females that escape predators mate. After the eggs have developed, the female lights on the stem of a cattail or other water plant. She backs down the stem until she is under water. Then, protected by a film of air around her, she uses the ovipositor to drill a double row of tiny holes in the plant's stem. These are the "nests" into which she thrusts her eggs.

Some three weeks later the eggs hatch into tiny long-legged nymphs. A nymph breathes through gill slits, back of which are twelve rows of thin gills. The gills strain oxygen from the incoming water.

To expel the water the nymph contracts the walls of its stomach. This contraction forces the water out through the gill slits. And at the same time the nymph is shot forward by the outgoing water.

A nymph moves about underwater in this manner, looking for insects to feed on. There is an odd jointed section on the under lip, or labium, that has two hooks. This section folds over the mouth parts. But when prey comes too close, the nymph thrusts out this hooked section to grasp the insects on which it feeds.

In the nymph stage the dragonfly moults perhaps as many as twelve times. After each moult the creature is longer and the wing covers have increased in length, too. But before a nymph makes its final moult or moults, the insect has to undergo a dormant stage.

The nymph of the green dragonfly becomes inactive by stages. Early in the winter it goes to the bottom of a pond only at night or on dark days. But as long as the sun is still strong enough to make the water warm at midday, the nymph climbs back up a stem until it is near the surface.

When the water remains cold for twenty-four hours at a time or is sealed by a lasting coat of ice, a nymph lapses into a dormant state in the mud at the bottom of the pond.

BUMBLEBEE

In summer a northern forest is a noisy place. But in winter, when ice stills the gurgle of streams, the same forest is apt to be quiet. The silence at this season is due to the fact that many of the forest's wild creatures are deep in winter sleep.

One of these deep sleepers is the young queen bumblebee. She is the only member of the colony that lives through the winter. To pass the season in safety a queen tucks herself into a mossy stream bank or secretes herself deep in a rocky crevice.

In these or other spots to which frost does not penetrate a queen bumblebee passes the winter months. As she hibernates, the rate at which her energy is produced slows until it reaches a point that barely sustains life. And then in a seemingly deathlike state the queen remains until the warmth of spring awakens her and she assumes the duties that produce a new colony.

Not all queen bumblebees awaken at the same time. Among the 5,000 species of wild bees in the United States and southern Canada there are several races of bumblebees. They have diverse habits and habitats. Some queen bumblebees emerge from their wintering spots early in April; others do not appear until late in May.

The first of each season are often out at the time when the catkins of the pussy willow are yellow-gold with pollen. A recently awakened queen is still sleepy; she crawls slowly back and forth over the catkins. As she meanders her "baskets" fill with grains of pollen.

These baskets are not the kind you and I are familiar with. Two sections on each hind leg of the bee are edged with long, stiff, and inward-growing hairs. These hairs catch and hold the pollen as the bee moves about on one catkin after another.

As soon as the queen is wide awake and fed, she searches for a good spot in which to build a nest. She usually selects a site on the ground, and frequently builds her nest of sphagnum moss. This moss grows in water or wet places. Nests are also built among grasses or even in the runways of meadow mice.

Nests are constructed underground, too. Usually these homes are in soil that packs firmly. Depending upon the species of bumblebee, such nests may be at the end of a 6-inch tunnel or at the end of one that is as long as 9 feet.

When a queen decides on a site, she begins the work that is to start a new colony of bumblebees. A colony will include worker bees, other queens, and males.

First, the queen prepares a small bed of woolly material. Next, she constructs one or more waxen cells—somewhat rounded and arranged in an irregular pattern. Finally, she molds a cup of wax, which she fills with honey.

Honey is made from the nectar a queen gathers from various flowers. She is able to reach certain nectars that other species of bees cannot even touch. Her ability to do this is because the bumblebee belongs to one of four families of wild bees with tongues so long that they are known as the long-tongued bees.

Among these families the bumblebee has one of the longest tongues. Its length is such that the nectar of even the red clover is within easy reach. Red-clover nectar is contained in the flower envelope. This little pocket is among the petals and at the very end of a long tube.

The queen gathers nectar until the wax cup is full. Then she lays 4 to 8 eggs. Her attitude toward the eggs and the young that hatch is like that of a brooding mother bird. After the eggs hatch, she feeds the young bees honey and increases the pollen supply if necessary.

The bees in the first hatching are all workers, all females, and in size all are small. Soon after emerging from the nest

Bumblebee on pussy willow.

the workers start to gather nectar and pollen from various wild flowers. Each day they make numerous round trips between the hive and the source of supply. They also do all the work around the hive. This relieves the queen of duties of any kind; the rest of her life is devoted to laying eggs.

After the first brood or two, newly born bees are larger at birth than their predecessors. There are more workers now to gather nectar and pollen and more workers to feed the young. By midsummer many of the newly hatched bees are fed all they can eat. Most of these especially fed bees develop into queens.

At this season males begin to appear in the colony. The number of bumblebees in a colony varies. There may be as few as 50 or as many as 200. And the make-up of the colony is a mixture of all types.

The males are known as drones. These bees are lazy individuals who take their ease in the flowers of goldenrod. As well as being loafers, they are also great drinkers. They consume quantities of the nectar collected by the worker bees.

A queen mates with a male outside the nest. After mating, a young queen has a few days or weeks of freedom. During this period she is seldom in the nest and often in the air.

As summer comes to a close the old queen bumblebees stop laying eggs. The colony becomes smaller and smaller as these and its other members die. Then the time arrives when only the young queens are left.

Young queen bumblebees go into hibernation over an extended period. The species and the whereabouts of the range are factors that influence the time of hibernation. Some are gone from the woods and fields by mid-July. Others are still flying until mid-October.

But whether hibernation is early or late, the queen bumblebees lapse into a deathlike sleep. And while they sleep the nests that they built the previous spring are destroyed by the larvae of moths and beetles.

Mollusks

The animals classed as mollusks are usually distinguished by a soft body that is protected by a hard shell. The shell, of their own manufacture, is of a lime-like consistency. Many members of this class have a spurred shell, and in some ways they are more advanced than any animal group with the exception of the vertebrates.

Mollusks live in the sea, or in fresh water, or on land. Those living in water attach themselves to rocks and piles, bury themselves in the sand, or swim freely through the water. Land forms spend a great part of their lives under rocks and stones and in other places of concealment.

GARDEN SNAIL

At the approach of winter the garden snail tucks itself into a safe spot. It crawls into fissures of rocks, or under heaps of stones and piles of dead leaves, or buries itself well beneath the surface of the earth.

In these locations the animal, whose shell is brown in color and spiral in form, is not exposed to the killing bite of winter.

After a few days of spring sunshine the snail emerges from its winter hideaway to look for food and a mate. As the animal inches along over a threadlike course, it leaves a trail of silvery slime.

A trail shows plainly when the ground is rain-drenched or sopping wet from dew. The snail covers its territory on a single, long, flat "foot." This foot is a part of the belly, and the snail uses it to coat a path with a sticky substance. Then parts of the foot grip the ground, while other parts move forward.

Thus the snail travels by drawing or pulling itself along. This means of locomotion is known as "traction." And by drawing and pulling 35 times a minute, the snail moves ahead about 2½ inches in a minute.

To locate its food the garden snail uses its highly developed senses of touch and smell. This animal eats all kinds of tender plants, though certain snails eat animal as well as vegetable foods. Some species are garden pests, but more damage to flower and vegetable plants is done by the slug—another hibernating mollusk.

Some little time after emerging from hibernation snails mate. Once this has been accomplished, the animals seek places in which to lay their eggs. A small hollow in loose soil is usually the "nest" in which snails leave their eggs. The egg-laying may take two days and when it is all over the weight of the eggs often equals one third of the weight of the adult snail.

A year after the snail is hatched it reaches maturity. It may live as long as five years, provided that each fall it goes into hibernation in a healthy condition. During the pre-hibernation stage, the snail eats quantities of food. At this time, too, many hard bits of lime form in the flesh.

When the snail settles down in its wintering spot, the

Garden snail.
This shows the shell turned on its side in order to show plainly the stopper the snail builds to plug the entrance.

bits of lime are used to build a weatherproof door. The limy particles are secreted and mixed with mucus to form a stopper for the opening of the shell.

Then, hidden in a spot to which the frost will not penetrate and with the door closed tightly to keep out the cold, the snail lies insensible to winter weather.

Not all land snails hibernate. Desert forms often have a long-lasting summer sleep. And one has something of a record among estivating animals.

To avoid dry conditions this snail passed more than four years in a dormant state. When it finally "came back to life," a staid museum staff was practically stupefied.

The snail that slept so long was from Egypt. It was in a collection of fauna shipped from that country to a museum in England. In March, 1846, someone at the museum thought that the snail shell was empty and mounted it on an identification card.

Four years later, March, 1850, traces of slime were evident on the card. Card and shell were quickly doused in water. When the shell came off the card, presto! out crawled a snail.

A little later someone brought a lettuce leaf to the snail. It started to eat at once. Thus the prolonged sleep and fast of the Egyptian desert snail came to an end in the arid atmosphere of a museum, more than 2,200 air miles from the arid habitat that was the point of origin.

In Conclusion

No ANIMAL in North America can match the record for prolonged sleep set by the Egyptian desert snail. But one insect has a childhood that is outstanding due to its length. This is the animal most of us call the seventeen-year locust but which scientists have named the periodical cicada.

Seventeen years after the eggs of the periodical cicada hatch in the soil beneath a tree the brood emerges. Sometimes there are as many as 40,000 in a single brood. The holes out of which they come are close together—frequently so close that 84 may perforate one square foot in the surface of the soil around a large tree.

The four-year sleep of the Egyptian desert snail and the seventeen-year childhood of the periodical cicada are extremes in suspended animation and slow growth. Most animals behave in a much less spectacular manner.

We have told you a lot about various hibernating mammals in North America—those that overwinter in a deathlike state. We have also written about the winter sleep of bears and three of the smaller flesh-eating mammals. The body functions of these animals slow but little or not at all during their inactive periods.

Other North American species meet winter in different ways. Some migrate to another part of this continent or to another part of the world. Most migratory creatures are birds. A few others, however, travel long distances to reach warmer climates. One of these is the monarch butterfly, distinguished by a double row of white dots in the black border of its orange wings. It flies from Ontario, Canada, all the way to Florida, to southern Texas, and to California. Another migrant butterfly, the cloudless sulphur, flies across the Caribbean to Bermuda.

North America's migratory birds generally fly to Florida, Cuba, or Central and South America. But some that summer on this continent go even farther.

From its summer home in Alaska the Pacific golden plover flies nonstop to its winter home in Hawaii. Some plovers do not even stop here. They go on to the Marquesas or other islands in the South Pacific.

In these faraway wintering areas migratory birds find the conditions that they need in order to live—a suitable climate in regions that afford protection and plenty of food.

Some animals stay at home all year round. Stay-at-homes are known as resident animals and they remain active during winter. A few of the continent's animals make local migrations. The mule deer of the West is one of these. Its migration is both local and vertical; in the fall it moves from its mountain pastures to more protected areas at lower altitudes.

Most resident animals stay in the same area for all twelve months, and in winter live much as they do in summer. But all summer long, usually a season when natural foods are abundant, they have been putting on fat by eating as much as they can. In the lean winter months, when food is harder to find, or even scarce, this extra fat is absorbed into their systems and helps to keep them well.

The white-tailed deer, for instance, acquires a layer of fat under almost all parts of the skin, with thicker masses on the hips and saddle. And with sufficient food and protected spots in which to rest, the animal is likely to come through the winter in good condition.

Some animals prepare for winter by acquiring denser fur on certain parts of the body. Others meet the season by undergoing a color change in the fur. In winter the soles on the feet of the forest-dwelling marten are furred. This seasonal covering protects the feet from ice and snow; it is a good substitute for snowshoes. The furred feet of the sleek little animal permit it to move across the surface of soft, fluffy snow without sinking in.

In winter the coat of the varying hare changes from summer brown to pure white. Squatted down in the snow, the varying hare becomes almost invisible due to its coat of winter white. This blending with the landscape makes it possible for many hares to escape numerous predators. For safety and also for warmth the varying hare frequently burrows into drifts of snow.

Hibernation and winter sleep, local and long-distance migration, and adaptation to the season are the ways in which most animals get through the winter. This is never an easy season for wildlife. For those that are active it is a period when each has to withstand cold and stormy weather as best it can. It is also a time when the active animal has to get enough to eat and to escape predators hunting for a meal in a season of scarcity.

Our hibernating and winter-sleeping animals seem to have the advantage over other species. If they have retreats that afford the correct temperatures and protection from the cold and from their enemies, they can survive—provided that they are in good condition at the onset of winter.

When winter is over these animals emerge from their hideaways. At first some are in a state of lethargy induced either by hibernation or winter sleep. But once fully awake they quickly resume their active roles in the various plant and animal communities throughout North America.

List of Common and Scientific Names

COMMON NAME	SCIENTIFIC NAME
Bat, little brown	*Myotis lucifugus*
Bear, black	*Euarctos americanus*
Beetle, convergent lady	*Hippodamia convergens*
Bumblebee	*Bombus americanus*
Butterfly, cabbage white	*Pieris rapae*
Carp	*Cyprinus carpio*
Caterpillar, *see* Woolly bear	
Chipmunk, eastern	*Tamias striatus*
Dragonfly	*Libellula pulchella*
European mantis	*Mantis religiosa*
Grasshopper	*Melanoplus femur-rubrum*
Hummingbird, ruby-throated	*Archilochus colubria*
Mackerel, fork-tailed	*Scomber scombrus*
Marmot:	
Yellow-bellied	*Marmota flaviventris*
Hoary	*M. caligata*
Moth, Isabella tiger	*Isia isabella*
Mouse, jumping:	
Grassland	*Napaeozapus insignis*
Woodland	*Zapus hudsonius* and *Z. princeps*
Poor-will	*Phalaenoptilus nuttalii*
Prairie dog:	
Black-tailed	*Cynomys ludovicianus*
White-tailed	*C. leucurus*
Raccoon	*Procyon lotor*

Common Name	Scientific Name
Shark, basking	*Cetorhinus maximus*
Skunk	*Mephitis mephitis*
Snail, garden	*Helix hortensis*
Snake, garter	*Thamnophis sirtalis*
Spider, black and yellow garden	*Miranda* (Argiope) *aurantia*
Spring peeper	*Hyla crucifer*
Squirrel, Columbian ground	*Citellus columbianus*
Toad	*Bufo americanus*
Turtle, box (land tortoise)	*Terrapene carolina*
Turtle, snapping	*Chelydra serpentina*
Woodchuck	*Marmota monax*
Woolly bear (caterpillar)	*Isia isabella*

ABOUT THE AUTHOR

Will Barker writes that since his high school days he has been primarily interested in two subjects—nature and writing. He has turned this avocation into a vocation, and is well known for his books and articles in the field of natural history.

The author of FAMILIAR ANIMALS OF AMERICA, Will Barker has also written many articles for various magazines, including *Natural History, American Forests,* and the *American Junior Red Cross News.* He has been cited by the National Wildlife Federation for his work in conservation. Mr. Barker lives in Washington, D. C.

ABOUT THE ARTIST

Carl Burger's love of the out-of-doors has been growing since his boyhood in Tennessee. A graduate of Cornell University, Mr. Burger has illustrated many books for children and adults, including Will Barker's FAMILIAR ANIMALS OF AMERICA and OLD YELLER by Fred Gipson. Mr. Burger and his wife live in Pleasantville, New York.

LOMA VISTA
P.T.A. LIBRARY